W9-ACV-800

RATIONAL CONCLUSIONS

JAMES D. AGRESTI

DOCUMENTARY PRESS

Rational Conclusions
Copyright © 2009 by James D. Agresti

www.RationalConclusions.com

Published by Documentary Press

Library of Congress Control Number: 2009940070
ISBN 13: 978-0-615-33236-9
ISBN 10: 0615332366

Unless otherwise stated, all Biblical quotations are from the King James Version.
Typographical corrections and minor changes incorporated October 4, 2017

Manufactured in the United States

CONTENTS

Preface VII

Introduction IX

Chapter 1 | Ancient History 1

Chapter 2 | Science 35

Chapter 3 | Archaeology 53

Chapter 4 | Continuity. 91

Chapter 5 | Ancient Manuscripts, Modern Bibles, and
       The Da Vinci Code 119

Chapter 6 | Hostile Witnesses, Cosmology,
       and Biogenesis. 141

Chapter 7 | Genetics 171

Chapter 8 | Embryology and Vestigial Organs 199

Chapter 9 | The Fossil Record. 233

Chapter 10 | Authentic Christianity 303

Appendix | Details on Frequently Cited Sources 321

Citations and Index . . . . www.RationalConclusions.com

PREFACE

When it comes to reading literature that is supposed to be informative, the only thing more frustrating than finding less information than we need is being swamped with more than we want. To remedy this typical conflict between thoroughness and brevity, this book was written in a special format designed for readers with varied levels of interest.

The main text provides the key points and details, which should be ample enough to satisfy the vast majority of people. The sections of text enclosed by arrows and printed in a contrasting font ▶like this◀ offer additional materials for those who would like to explore certain topics in greater depth. These are primarily for academically minded readers and can be skipped without missing crucial information or disrupting the flow of the book.

To establish the accuracy of the hundreds of facts that appear in the forthcoming pages, I have documented them far more thoroughly than standard academic practice requires. Over years of conducting research on diverse subjects, I cannot begin to count the number of times I have looked up a source citation and found it was misrepresented. Therefore, as is possible without infringing upon copyright laws that limit the amount of text I can reproduce from any given source, the citations of this book directly quote the sources so readers can verify they are accurately represented. Instead of printing these comprehensive citations in this book, I have posted them online at rationalconclusions.com to keep the physical size of this book to a minimum (as it would more than double in size if this material were among the pages you are holding) and to offer clickable access to sources available online.

An important point to understand about the sources cited in this book is that they are predominantly technical periodicals, scholarly journals, reference works such as encyclopedias, and books found in the science and history sections of university libraries. One of the principal features that make this book unique is the exhaustive use of highly credible sources. My objective was to document all key facts so thoroughly that no reasonable person would doubt them. These citations are also available in e-reader versions of this work.

Notwithstanding the academic source materials, this book was written so it could be understood without a specialized education in any of the topics examined. To accomplish this, brackets [ ] are used in a slightly unconventional manner to replace technical verbiage with plain language that has the same meaning. For instance, in scholarly literature, the first five books of the Bible are often referred to as "the Pentateuch." Hence, when quoting from a source that uses this term, I have replaced it with "[the first five books of the Bible]." For those who would like to read the original wording, the online citations display the exact text and the plain language in brackets. Thus, brackets [ ] are used to paraphrase; parentheses ( ) contain the exact wording of the original sources, and curly brackets { } are used to enclose my own comments and thoughts.

Traditional academic citations, which are generally rife with abbreviations, are dispensed with in favor of a system that spells out almost everything. Also, many academic works hail from countries like England where certain words are spelled differently from U.S. spellings (for example, "colour" versus "color"). Thus, for the sake of readability and with apologies to our friends across the Atlantic, I have maintained the Queen's English in the titles when citing these works but have debased them with American spellings when quoting from the main bodies of the texts.

Note that throughout this book and the online citations that support it, I have not *italicized* or **emboldened** anyone else's words. Where you see such emphases, they are those of the original authors.

Sincere thanks to the following individuals for their vital contributions to this work or for supporting me during this ten-year endeavor: Gina Diorio, Stephanie Spino, James & Janet Agresti, Joe Agresti, Tom Agresti, Fran and Christine Neville, Ed Quinn, Steve & Melissa Mitas, Steve & Julia Cardone, Frank & Audrey Cann, Jean-Marc & Katerina Kopp, Alice Senior, Dave & Amy Lee, Nicole Armbruster, Erik Codrington, Jon Heller, David Dew, Ken Gayer, and the many learned scholars and researchers who laid the informational base for this book and whose shoulders I stand upon. Finally, I would like to thank my wife and children, who have willingly sacrificed so much to make this work possible.

INTRODUCTION

How do we establish our spiritual beliefs? Many follow in the footsteps of their parents or other influential figures in their lives. Some embrace their views on the basis of emotional appeal. Others invest a certain amount of thought into the process but often do so without performing substantive research. If you were to ask the people you know why they believe what they do, how many do you think could give you a rational answer? For many people, spiritual beliefs are a matter of personal preference or blind faith.

The great irony here is most people would agree spiritual beliefs impact our lives in significant ways, and many, including me, think they have eternal importance. Why then would anyone entrust the formation or rejection of such views to whim or speculation? Given what is at stake, shouldn't careful investigation and serious thought be a part of the process?

The purpose of this book is to examine facts that can be used to arrive at rational conclusions regarding the Bible. Surprisingly, many of these facts proceed from academic disciplines such as:

Genetics
History
Archaeology
Paleontology
Physics
Cosmology
Embryology
Neurobiology
Microbiology

In the realm of spirituality, one of the easiest things to do is make simplified and unsupported assertions that are accepted by people who share the same mindset. The real test for any work that stakes a claim to truthfulness, however, is whether or not it can withstand the scrutiny of a judicious audience.

Hence, this is not a book for those who uncritically accept what they want to believe and robotically deny what they don't. It is for people who ask, "How do you know that?" and then follow up by asking, "How do you know that you know that?" Legitimate answers to such questions do not typically make for leisurely reading material, but the alternative of blindly embracing that which appeals to our notions or emotions is woefully inadequate for an issue of such magnitude.

CHAPTER 1
ANCIENT HISTORY

I used to think the people described in the Bible were legendary figures who were no more real than the gods of Greek mythology. This was primarily because I never read or heard anything about these Biblical characters from a historical perspective. Although not a history buff, I was still exposed to a reasonable amount of history through school, television, books, newspapers, magazines, conversations, and travel. Yet, in all these venues and even in the church I sometimes attended as a child, the events of the Bible never seemed to be discussed in a tangible manner—if at all. In fact, listening to some people talk about characters and situations in television sitcoms, I often felt a greater sense of realism than listening to anything I ever heard about the Bible. All of this left me with the impression that the Bible had no basis in reality.

When I took the time to critically examine this belief, however, I realized it was untenable given that numerous people, places, and events detailed in the Bible are corroborated by creditable independent sources. In this chapter, we will examine various ancient Roman and Jewish writings that contain such particulars. Before we do so, however, it is important to understand a few general points about ancient historical writings that will help us to understand the significance of what follows.

ESSENTIALS OF ANCIENT HISTORY

As simple and obvious as this may seem, it should be recognized that the objective of studying history is to find out what truly happened in the past. However, many people's knowledge of ancient history consists of what they have learned from modern authors and commentators. This can present a major problem because these modern commentators have the opportunity to distort and filter what the original sources have written —thus, the term "revisionist history." Yet, even reading the ancient sources is no guarantee we will receive accurate information. Just because someone wrote something down a long time ago doesn't necessarily mean it is true. Moreover, as we will see, the process of discovering what was actually written in ancient times can be fraught with uncertainty. In sum, despite the impression we may have developed from popular books or high school history classes, there are few simple answers in this field. With this in mind, let's begin our inquiry.

Up through the second century A.D., literature was primarily written on material called papyrus, which is made from a plant of the same name. This paper-like product was rolled around wooden cylinders to form scrolls. Papyrus is very vulnerable to moisture and deteriorates quickly when handled.[1] [2] Therefore, unless documents were deposited in a safe and dry location, they perished. Most ancient papyrus documents in existence today were discovered in Egyptian garbage dumps where vast piles of waste shielded the lower levels from rain and moisture. Another important source of papyrus documents is mummies, which were sometimes layered with used manuscripts.[3]

Papyrus manuscript dated to the second century A.D.[4]

Papyrus was eventually replaced by parchment, which is made from animal skins and is quite durable. By the fourth century A.D., parchment had become the dominant writing material, and some of the oldest books in existence today are made of parchment and date to this era.[5] [6]

In addition to the practical difficulties of preserving ancient documents, centuries of wars, fires, theft, floods, hurricanes, and other man-made and natural disasters have caused mass extinctions of such manuscripts. We can't be exactly certain how many works have been lost over time, but because ancient historians referenced the writings of their contemporaries, we are aware of many works that once existed and are now lost.[7] For instance, it is estimated that about 90% of the major Greek classics are permanently lost.[8]

Generally speaking, in the field of ancient history, there is no such thing as an original. By that, I mean a manuscript that was written by the original author. Given the high probability that any particular manuscript would be destroyed, the only way an ancient work usually could survive is if it were copied by hand. Remember, there were no printing presses, copy machines, or Internet. Imagine trying to manually copy the contents of an entire book or volume of books. Do you think you would make mistakes? By comparing different copies of the same work, we have learned that errors were common in this process. Furthermore, since the reproduction of writings was such an arduous process, only a small percentage of ancient works even made it into circulation in the first place.[9] [10]

With regard to accuracy, we must also consider the obstacles authors and copyists faced in ancient times. For the most part, there were no indexes, page numbers, punctuation, or even spaces between adjoining words.[11] [12] [13] Additionally, the tools we now use for research were either scarce or non-existent. Even now, research can be an arduous process, so imagine trying to conduct it without the benefit of modern-day libraries and computers. Also, just as today, what we can establish about a writer's meticulousness and honesty all plays a role in how believable we consider the writer to be.

Some try to exploit the realities that are common to ancient history by selectively demanding unreasonable standards of proof for any evidence that runs contrary to their beliefs. For instance, there are those who claim certain historical works cannot be relied upon because large gaps of time exist between when they were originally written and the oldest surviving copies. Yet, gaps of many centuries are typical when dealing with works that are 2,000 years old. In fact, the bulk of ancient Greek and Latin writings that exist today are known through copies that were made between the ninth and fifteenth centuries A.D.[14] For example, Julius Caesar's *Gallic War* was written in the middle of the first century B.C., yet the oldest existing copies date to the ninth century.[15] Other historical works that are regularly cited and relied upon, such as those of Thucydides and Polybius, have gaps of a millennium or more between when they were written and the oldest known copies.[16] [17]

The bottom line is that if we are going to examine the Bible by comparing it to other ancient works, we should temper our conclusions with the realization that there is room for uncertainty. On the other hand, there comes a point at which enough concurring evidence has been accumulated that it would be irrational to deny the reality of certain events.

Now that we have established a basic appreciation for the nature of ancient historical evidence, let's examine some of it.

The Greatest Historian of the Roman Empire

Cornelius Tacitus was born in about 57 A.D.[18] He held several important positions in Roman government, including the most prestigious office a private citizen could hold, that of a consulship.[19] [20] He is commonly regarded as the "greatest historian" of the Roman Empire.[21] [22] At about 60 years of age in 117 A.D., Tacitus completed a work called the *Annals* in which he describes the following events that occurred in 64 A.D. during the reign of the emperor Nero:[23] [24] [25]

> Consequently, to get rid of the report, Nero fastened the guilt and inflicted the most exquisite tortures on a class hated for their abominations, called Christians by the populace. Christus, from whom the name had its origin, suffered the extreme penalty during the reign of Tiberius at the hands of one of our procurators, Pontius Pilatus, and a most mischievous superstition, thus checked for the moment, again broke out not only in Judaea, the first source of the evil, but even in Rome, where all things hideous and shameful from every part of the world find their center and become popular. {The forthcoming sections printed in a different font and enclosed by arrows are intended for academic readers.} ▶Accordingly, an arrest was first made of all who pleaded guilty; then, upon their information, an immense multitude was convicted, not so much of the crime of firing the city, as of hatred against mankind. Mockery of every sort was added to their deaths. Covered with the skins of beasts, they were torn by dogs and perished, or

were nailed to crosses, or were doomed to the flames and burnt, to serve as a nightly illumination, when daylight had expired. Nero offered his gardens for the spectacle, and was exhibiting a show in the circus, while he mingled with the people in the dress of a charioteer or stood aloft on a car. Hence, even for criminals who deserved extreme and exemplary punishment, there arose a feeling of compassion; for it was not, as it seemed, for the public good, but to glut one man's cruelty, that they were being destroyed.[26] ◄

If this account is credible, Tacitus corroborates the existence of Christ and affirms he was executed by Pontius Pilate, just as detailed in the following Biblical passages:

And Jacob begat Joseph the husband of Mary, of whom was born Jesus, who is called Christ.[27]

And so Pilate, willing to content the people, released Barabbas unto them, and delivered Jesus, when he had scourged him, to be crucified.[28]

Tacitus also asserts Tiberius was the emperor of the Roman Empire at the time, as does the Bible:

Now in the fifteenth year of the reign of Tiberius Caesar, Pontius Pilate being governor of Judaea.... And Jesus himself began to be about thirty years of age....[29]

We also find concurrence between Tacitus and the Bible with regard to Judea being the birthplace of Christianity. The following is one of many Biblical statements that set the origin of Christianity in the cities and towns of Judea:

And [Jesus] led them out as far as to Bethany.... And they worshipped him, and returned to Jerusalem with great joy.[30] [31]

The oldest existing copy of Tacitus' book quoted above dates to the 11th or 12th century.[32] [33] There are 32 manuscripts of it, and this passage appears in all of them.[34] [35] However, all but one of the newer manuscripts stem from the oldest one,[36] which means they do not provide independent confirmation of one another. The remaining manuscript may derive from a closely related predecessor to the oldest, or it may stem from the oldest as the others do.[37] Hence, it does not offer much in the way of independent confirmation either.

So how do we know that Tacitus wrote what is quoted above or, for that matter, that he even existed? First, since 100 A.D., writers in every century except for the seventh and eighth have mentioned or quoted him.[38] Also, the specific passage we are examining fits the context of the surrounding narrative, and the verbiage in it is characteristic of Tacitus' other writings.[39] Additionally, since this passage refers to Christianity as an "evil" and states Christians were "criminals who deserved extreme and exemplary punishment," it is highly unlikely that a Christian copyist simply inserted it at some point in the distant past. Thus, there is no rational reason to doubt this passage is authentic.

The next step in evaluating this passage is to determine how much importance should be accorded to it. This is not only good academic practice, but it is also especially important in this case because it is generally accepted that Tacitus wrote the passage, yet some still argue it doesn't prove much insofar as Jesus is concerned. The primary basis for this is the claim that Tacitus relied upon hearsay from Christians for this information. If such is true, this quote would be inconsequential because it would simply amount to the recycling of Christian claims.

Critical to conducting accurate research is understanding the opposing arguments that seem to beset many matters of importance. However, much time can be wasted in evaluating frivolous claims, and there is no need to overload this book with a refutation of every baseless argument in existence. Instead, I will debunk a cross-section of such claims to illustrate the types of flaws from which they suffer and leave it to the reader to take it from there. The first order of business is to examine arguments behind the claim that Tacitus obtained his information from Christians. One such argument is as follows:

> [H]e gives Pilate a title, procurator, which was current only from the *second* half of the first century. Had he consulted archives which recorded earlier events, he would surely have found Pilate there designated by his correct title, prefect.[40]

This assertion is faulty on several accounts. First, there is a distinct possibility Pilate was both a prefect and procurator. An inscription discovered in 1961 refers to him as a "prefect";[41] Tacitus and the Jewish historian Josephus call him a "procurator,"[42 43 44] and most tellingly, the Jewish historian Philo (a contemporary of Pilate[45 46]) refers to him as "one of the prefects appointed procurator of Judea."[47] Furthermore, several other Roman officials were referred to by both titles.[48 49] Shockingly, the academic who put forward the objection above is well aware of all this, but very few of his readers are because he buries this information in a footnote of an article on the Internet while providing not the slightest hint in the main body of the article that he is making any sort of concession.[50]

Furthermore, the notion that Tacitus, a high-ranking Roman official, would depend upon Christians to inform him about the title of a Roman governor is preposterous. Why would Tacitus trust people he calls "criminals" to educate him in his own field of expertise? As explained in a creditable modern work called *The Cambridge Ancient History*, the offices Tacitus held endowed him with:

> an advantage denied to many of the historians mentioned here, of knowing the workings of the system he was to describe.[51]

Most importantly, this entire argument is founded upon the irrational assumption that Tacitus would duplicate the exact wording used by his sources. It was and still is a common practice for authors to write in the vernacular of their audience. If "procurator" were the title for the governor of a Roman province in Tacitus' day, it would make perfect sense for him to use this term even if his source or sources did not. And it just so happens that Tacitus wrote in precisely this type of manner, using contemporary verbiage and shunning technicalities so that his audience would find his writings to be understandable.[52]

▶Here is another example of specious logic put forward to argue Tacitus was "simply repeating what Christians had told him":

> Tacitus does not name the executed man Jesus, but uses the title Christ (Messiah) as if it were a proper name. But he could hardly have found in archives a statement such as "the Messiah was executed this morning.[53]

First, there is no legitimate reason for replacing "Christ" with "Messiah" in this context. Messiah is a Hebrew word that means "anointed," and Christ is the Greek version of this Hebrew word.[54] But Tacitus didn't write in Hebrew or Greek. He wrote in Latin, and we are reading him in English more than 1800 years later. Given the fact that the meanings and usages of words change as they are passed across languages, cultures, and time, it is misleading to substitute "Messiah" for "Christ" in this context. For example, in English today, it is perfectly normal to use the word Christ as a proper name. We say, "Christ was executed under Pontius Pilate."

So let's move on to analyzing what Tacitus wrote instead of what someone has tried to put into his mouth. The key point is that in Latin and English, the word Christ is a transliteration,[55][56] or a word that is arrived at by taking letters from one language and changing them into the equivalent letters of another language.[57] This differs from a literal translation, in which the exact meaning of a word is carried over. When words are transliterated, the original meaning is often obscured. Again, the word Christ is a perfect example. Most people today have no idea that "Christ" literally means "anointed." Therefore, we cannot rationally impute a spiritual or reverent connotation of this word to Tacitus or his source(s).

Moreover, this argument is built upon the same irrational assumption that Tacitus would replicate the exact wording used by his sources. Again, if the word Christ were used as a proper name in Tacitus' day, it would make perfect sense for him to do the same, even if his source or sources did not.◄

Now, on to evaluating the historical value of this passage. Since Tacitus wrote that Christ was executed during the reign of Tiberius (14 to 37 A.D.[58]), and since Tacitus was born in about 57 A.D., it is important to identify the types of sources he used. This will help us determine whether or not he had access to reliable information. Tacitus specifically identifies some of his sources, and by examining these, we can acquire a good sense of his practices. For an even more significant analysis, we can look at sources he cites for our specific timeframe of interest, which is the reign of Tiberius:

- "Caius Plinius, the historian of the German wars,"
- "Tiberius' own speeches,"
- "writers and senators of the period,"
- "Greek historians,"
- "the daily register,"
- "the narrative of most of the best historians,"
- "the memoirs of the younger Agrippina, the mother of the emperor Nero,"
- "Historians of the time," and
- "the notes of the proceedings furnished to the Senate."[59]

All of these tell us Tacitus had access to a wide variety of credible sources for the period in which both he and the Bible place Jesus.

Next, we should try to determine if the statements Tacitus makes in this passage fall into his areas of expertise. Tacitus apparently had little interest in Christianity. Out of his existing writings, his only mention of Christianity comes in the passage we are evaluating. However, his primary field of expertise was Roman government, and his works abound with the names, titles, and actions of a vast number of Roman officials, from emperors down to soldiers. In fact, according to *The Cambridge Ancient History*, Tacitus' writings are "by far" the most comprehensive and reliable ancient source of information on the Roman government during the reign of Tiberius.[60] Considering that Pontius Pilate was the governor of a Roman province during this very period, there is no question his actions fall into Tacitus' area of expertise.

What of the possibility that Tacitus relied upon hearsay for his statements about Pilate and Christ? This is highly improbable for two reasons. First, Tacitus explicitly warns his readers not to accept hearsay:

> My object in mentioning and refuting this story is, by a conspicuous example, to put down hearsay, and to request all into whose hands my work shall come, not to catch eagerly at wild and improbable rumors in preference to genuine history which has not been perverted into romance.[61]

> So obscure are the greatest events, as some take for granted any hearsay, whatever its source, others turn truth into falsehood, and both errors find encouragement with posterity.[62]

Second, although it is possible someone could make statements like this and then totally disregard them, we know Tacitus did not because he made extensive use of qualifiers. For example, there are many instances in which Tacitus is not 100% sure of something and makes this clear to his readers. Consider this sampling from the first book of the *Annals*, which is only about 25 pages long:

- "some thought,"
- "Whatever the fact was,"
- "it has not been thoroughly ascertained,"
- "The common account is,"
- "according to some,"
- "Some have related that,"
- "it was said," and
- "I can hardly venture on any positive statement."[63]

It is worth noting that Tacitus uses no qualifiers in his explanation of the origin of Christianity:

> Christus, from whom the name had its origin, suffered the extreme penalty during the reign of Tiberius at the hands of one of our

procurators, Pontius Pilatus….

For a vivid contrast, we can look at the manner in which Tacitus writes about the origin of Judaism. In this case, he cites five different accounts and precedes each with a qualifier such as "Some say" and "Others assert."[64] Conversely, when Tacitus writes that Christ was executed by Pilate, he plainly states it as a fact.

Additional evidence that reflects well on Tacitus' credibility is that he makes a point of comparing multiple sources. In fact, he writes:

> Proposing as I do to follow the [concurring] testimony of historians, I shall give the differences in their narratives under the writers' names.[65]

He doesn't always employ this practice, but throughout the *Annals*, it is frequently clear that Tacitus has access to and makes use of more than one source to depict the same events:
  • "As to that point historians differ,"
  • "authors have given both accounts,"
  • "So far our accounts agree,"
  • "the narrative of most of the best historians,"
  • "There are various popular accounts,"
  • "writers of the time have declared," and
  • "It is related by several writers of the period."[66]

Finally, the last test for this passage is to evaluate Tacitus' track record for reliability. Towards this end, I consulted as many different scholarly works on the topic as I could find.[67] As the quotations below demonstrate, eight of the nine works I located articulated exceptionally high opinions of Tacitus' factual accuracy:

> He is by far the most complete and most trustworthy author that we possess for the early [Roman Empire].[68]

> Even his severest critics concede the general accuracy of the facts that he records. This can be proved … by checking many of his details with evidence derived from inscriptions, coins, and other archaeological material.[69]

> [The] infinite effort [of critics] has failed to produce evidence of false statements beyond those occasional mistakes which no mere

human can hope to escape.[70]

His passionate opinions should not obscure the fact that he is the most accurate of all the Roman historians.[71]

Tacitus can rarely be detected in falsehood, and by the standards of the ancient world he is careful and conscientious in his search for the truth.[72]

Tacitus provides an agreeable channel whereby the best historical insight and experience of the ancient world are made available for our own enlightenment....[73]

Cornelius Tacitus does not need to be vindicated for accuracy. He consulted a variety of sources, and he was at pains to establish the truth.[74]

Tacitus' presentation of the facts, where it can be checked, is generally as reliable as that of any ancient historian....[75]

Of course, nothing is ever simple, and one of the nine works I consulted offers a generally low opinion of Tacitus' accuracy, claiming he "makes many mistakes." The problem with this opinion is that the author uses an endnote to support it, and this endnote cites two sources, both of which do not substantiate his assertion. In fact, both sources are among the eight works quoted from above.[76]

Beyond the generalizations, I also took the approach of researching specific instances of error in Tacitus' writings. What I found was a small number of mistakes along the lines of claiming that the earth is flat and stating that a particular woman was the wife of the emperor Nero's grandfather when in fact it actually was the woman's older sister.[77] [78]

Although allegations of factual errors in Tacitus' works are few and far between, even some of the ones I came across are clearly unwarranted. For example, one of the scholars who has a generally high opinion of Tacitus faults him for "making the Jews natives of Crete."[79] The problem is that Tacitus does no such thing. What he does is relate that this is the belief of some people. He writes, "Some say that the Jews were fugitives from the island of Crete...." This appears in the passage mentioned earlier in which Tacitus contrasts five different accounts of the origin of the Jews. To reiterate, Tacitus precedes each and every account with a qualifier, making it abundantly clear he is uncertain about the facts of the matter.[80]

As the sweeping evidence above reveals, Tacitus' statements about

Christ and Pilate were written by an accomplished historian who used a wide array of credible sources for the relevant timeframe, was operating within his field of expertise, exhibited skepticism of hearsay, identified areas in which he was uncertain, and demonstrated a track record of reliability that is among the best of ancient historians. Furthermore, this evidence does not stand alone. It concurs with the next ancient source we will examine.

THE GREATEST JEWISH HISTORIAN OF ANTIQUITY

Flavius Josephus, born in 37 A.D., was a Jewish priest and government official.[81] He is touted by modern scholars as the "the most important Jewish historian of the premodern era"[82] and "the greatest Jewish historian of antiquity."[83] [84] At 56 years of age in about 93 A.D.,[85] he completed a famous work now known as the *Antiquities of the Jews*. In it, he writes:

> And so [the high priest] convened the judges of the [high court] and brought before them a man named James, the brother of Jesus who was called the Christ, and certain others. He accused them of having transgressed the law and delivered them up to be stoned.

▶ In full:

> Upon learning of the death of Festus, Caesar sent Albinus to Judea as procurator. The king removed Joseph from the high priesthood, and bestowed the succession to this office upon the son of Ananus, who was likewise called Ananus. It is said that the elder Ananus was extremely fortunate. For he had five sons, all of whom, after he himself had previously enjoyed the office for a very long period, became high priests of God—a thing that had never happened to any other of our high priests. The younger Ananus, who, as we have said, had been appointed to the high priesthood, was rash in his temper and unusually daring. He followed the school of the Sadducees, who are indeed more heartless than any other of the Jews, as I have already explained, when they sit in judgment. Possessed of such a character, Ananus thought that he had a favorable opportunity because Festus was dead and Albinus was still on the way. And so he convened the judges of the Sanhedrin [high court] and brought before them a man named James, the brother of Jesus who was called the Christ, and certain others. He accused them of having transgressed the law and delivered them up to be stoned. Those of the inhabitants of the city who were considered the most fair-minded and who were strict in observance of the law

were offended at this. They therefore secretly sent to King Agrippa urging him, for Ananus had not even been correct in his first step, in order him to desist from any further such actions. Certain of them even went to meet Albinus, who was on his way from Alexandria, and informed him that Ananus had no authority to convene the Sanhedrin without his consent.[86]◄

If this account is credible, it corroborates the existence of Jesus, his brother James, and the high priest Ananus, all of whom are mentioned in the Bible.[87]

With regard to evidence that bears upon the credibility of this account, we know Josephus existed because from the first century onward, writers in every century have mentioned Josephus or quoted from him.[88] The oldest manuscripts of the book containing this account date to the tenth century;[89] at least 11 Greek and 129 Latin manuscripts contain this portion of the book,[90][91] and significant research has failed to produce one manuscript in which this account is substantially different or absent.[92][93][94]

Furthermore, the *Antiquities* was originally written in Greek, and all known Latin manuscripts derive from a translation done in the sixth century.[95][96][97] Since the Greek and Latin manuscripts diverged at this point yet still corroborate each other with respect to this passage, strong evidence exists that that the passage has been reliably transmitted for at least the last 1500 years. Additionally, this passage fits the context of the surrounding narrative, and the verbiage in it is strongly characteristic of Josephus.[98][99] So, in brief, the evidence weighs very soundly in favor of the passage's authenticity.[100]

To assess how much importance should be accorded to this passage, we don't have to conduct a detailed analysis of Josephus' sources and historical methods because Josephus was in a unique position to garner the facts of this incident. From the context, we can determine that Josephus sets the execution of James in Jerusalem around 62 A.D.[101][102] From Josephus' autobiography, we can also determine that Josephus was about 25 years old at the time, lived in Jerusalem, and personally knew the people who executed James (the high priest Ananus and judges on the high court).[103][104] Josephus made a trip to Rome somewhere around this period, so it is possible that he was not physically present in the city when the event took place, but it clearly occurred within his personal sphere of influence.[105]

Additionally, the Bible claims that James was a leader of the Christian church at Jerusalem.[106] The fact that Josephus specifically identifies James, while referring to those who were executed with him as "certain others," corroborates the Bible with regard to James' prominence. Since Josephus was also a notable person living in Jerusalem at the same time, there is a distinct possibility they were familiar with each other and met face to face. Thus, considering all of the above, this is an exceptionally credible reference.

Josephus' *Antiquities* includes another passage about Jesus, which, in contrast, has been a source of justifiable controversy:

> About this time there lived Jesus, a wise man, if indeed one ought to call him a man. For he was one who wrought surprising feats and was a teacher of such people as accept the truth gladly. He won over many Jews and many of the Greeks. He was the Messiah. When Pilate, upon hearing him accused by men of the highest standing amongst us, had condemned him to be crucified, those that had in the first place come to love him did not give up their affection for him. On the third day he appeared to them restored to life, for the prophets of God had prophesied these and countless other marvelous things about him. And the tribe of Christians, so called after him, has still to this day not disappeared.[107]

This passage has been uncritically quoted in a number of Christian publications, but there are good reasons to conclude it has been altered since Josephus wrote it.

The first is that a Christian author named Origen Adamantius, who lived from about 185 to 254 A.D.,[108] wrote in two of his works that Josephus did not believe Jesus was the Christ.[109] [110] This is consistent with Josephus' reference to James, where he wrote Jesus was "called" the Christ. The second reason is that Josephus' writings are quite lengthy, and this passage is very short. If Josephus thought Jesus had risen from the dead and was the Messiah, it is doubtful he would have devoted so little space to him. Third, quotations of this passage that appear in a 10th century Arabic work and a 12th century Syriac work are more compatible with the view that Josephus did not accept Jesus to be the Messiah. The Arabic work states that Jesus was "perhaps the Messiah," and the Syriac work states that "he was thought to be the Messiah."[111]

A wealth of academic literature has been written about this passage. Most scholars agree that some form of it appeared in Josephus' original work but that it was altered during the copying process.[112] Hence, some use this conclusion in attempts to discredit ancient works that corroborate the Bible. They claim the writings of historians such as Josephus and Tacitus cannot be trusted because they have been passed down to us by Christians. What these critics inevitably fail to mention is that the vast majority of classical literature that exists today owes its survival to Christians. To disregard the ancient writings that were preserved by Christians is to disregard nearly all of the ancient literature that stems from this region of the world. We will see this is an undeniable fact, and one that belies the oft-repeated and false claim that Christians systematically censored and destroyed these writings.[113] [114] [115]

WHO PRESERVED THE ANCIENT CLASSICS?

Ancient classical literature is comprised of Greek and Latin writings dating from about 700 B.C. to 250 A.D.[116] [117] The regions from which these works sprung were all at one time or another part of the Roman Empire, which at its peak encompassed the lands of the New Testament and most of the Old Testament.[118] Between the second and fifth centuries, the Roman Empire was greatly weakened by factors that have been a source of heated debate.[119] More than 200 contributing causes have been proposed, including excessive taxation, population decline, privatization of government functions, bulging federal bureaucracies, military spending, welfare programs, moral decay, lead poisoning, and lack of quality artistic expression.[120] [121] It is well beyond the scope of our discussion to sort this out, but let it be said that some writings on the subject seem to convey more about the political, social, and religious leanings of their authors than they do about ancient history.

Whatever the reasons, the Roman Empire experienced a long period of decline that left it vulnerable to attack. In the fifth century, Barbarian invasions destroyed the western portion of it, causing the entire region to plummet into anarchy: tribes waged vicious wars against each other, civilized society disintegrated, and cities were reduced to ruins.[122] This devastation triggered a 500-year era that is sometimes referred to as the Dark Ages. Throughout this period and up into the 12th century, few besides monks pursued academics.[123] [124] To quote the college textbook *Scribes and Scholars, A Guide to the Transmission of Greek and Latin Literature*:

> [A]ll readers in the Middle Ages were more or less devout Christians....[125]

During a period in the eighth and ninth centuries, Christian monasteries, schools, and libraries, under the patronage of a king named Charlemagne (a Christian, if only in name[126]), placed intense effort into copying and preserving the ancient Latin writings.[127] [128] [129] If we examine the oldest manuscript of every ancient Latin classic that exists today, we find that more than 90% of them were copied during this era.[130]

Although the western portion of the Roman Empire was dismantled, the eastern part survived and became what is known as the Byzantine Empire. This civilization was highly advanced, very much Christian, and existed for more than a thousand years.[131] Greek was its primary language, and 75% of the ancient Greek classics that exist today come from Byzantine manuscripts.[132]

In addition to preserving the Greek and Latin classics, Christians also preserved Jewish writings that would otherwise have been lost.[133] The point, again, is that if we were to dismiss all of the ancient works preserved by Christians, there would be practically no classical literature left.

Nonetheless, if these early Christians weren't trustworthy copyists, it would make sense to be very suspicious of any and all works they preserved. This begs the question of their reliability. Again, citing *Scribes and Scholars*:

> [O]n the whole, the fidelity with which the classics are transmitted is remarkable.[134]

Another reputable source, *The Oxford Companion to Classical Literature*, agrees:

> [I]t is clear from a comparison of later manuscripts with early [papyrus documents] … there was no appreciable corruption during this time and that the quality of surviving texts was hardly impaired.[135]

It is also clear that not every copyist was trustworthy, but this hardly justifies placing the mass of ancient classics under a cloud of suspicion, especially since these writings are often subjected to rigorous tests to evaluate their credibility. Such tests include examining crucial passages for factual and verbal inconsistencies with the rest of the author's writings, probing the context for discontinuities, comparing multiple manuscripts of the same work, and scouring the text for verbiage that came into use after the time period in which the work was written.[136] In sum, to discredit portions of historical works because they were relevant to the copyists who preserved them is entirely illogical. It is only natural that people would take the time and effort to preserve that which was of interest to them.

ARGUMENTS FROM SILENCE

Some of the frailest theories in the field of ancient history are those based upon elements of silence in the historical record. For example, in attempting to make the case that Jesus was simply a myth, some people have compiled lists of first- and second-century writers who made no mention of him. Besides the fact that such lists include people whose writings are mostly lost,[137] they neglect to explain how each of these authors would have been in a position to know of Jesus. Once we examine who these authors were, where they lived, and what they wrote about, it is evident that almost all of them had no way of knowing or no cause to write about Jesus.[138]

One more relevant point on this matter should be addressed; that of a Jewish author named Philo. As opposed to the other ancient writers that didn't mention Jesus, it would be reasonable to expect that Philo might. Therefore, he is frequently singled out. Consider this classic case in point:

> Philo was born before the beginning of the Christian era, and lived until long after the reputed death of Christ. He wrote an account of the Jews covering the entire time that Christ is said to have existed on earth. He was living in or near Jerusalem when Christ's miraculous birth and the Herodian massacre occurred. He was there when Christ made his triumphal entry into Jerusalem. He was there when the crucifixion with its attendant earthquake, supernatural darkness, and resurrection of the dead took place —when Christ himself rose from the dead, and in the presence of many witnesses ascended into heaven.[139]

This is clearly ridiculous, as there is no factual basis on which to stake the claim that Philo lived "in or near Jerusalem" or to justify chirping, "He was there…. He was there…." In his existing writings, Philo mentions visiting Jerusalem on one occasion,[140] and as far as it is known, Philo lived in Alexandria, Egypt, his entire life.[141] [142] There are more than 300 miles between Alexandria and Jerusalem,[143] and according to most historians, Philo didn't have access to a car, train line, or local airport.

For a vivid illustration of just how meaningless such arguments from silence are, let's consider the individuals who were the high priests of Judea during Philo's adulthood, which began no later than 5 A.D. and lasted until sometime after 40 A.D.[144] Josephus names seven consecutive occupants of this office who can be firmly dated to this period: ▶Ananus, the son of Seth (6–15 A.D.); Ishmaël, the son of Phabi (15–16); Eleazar, the son of

Ananus (16–17); Simon, the son of Camith (17–18); Joseph Caiaphas (18–36); Jonathan, the son of Ananus (36–37), and Theophilus, the son of Ananus (37–41).[145] ◄ Two of these people also appear in the Bible, and the family to which four of them belonged is mentioned in an ancient Jewish work called the Babylonian Talmud.[146] [147] Yet, none of these people appear anywhere in Philo's writings.[148] Based upon this, are we supposed to believe they never existed? That would be absurd. The same applies even more so to Jesus, as no source portrays him as someone who was more well-known than a high priest of Judea, which was the preeminent religious office in all of Judaism.[149]

Ironically enough, even if Philo did mention Jesus, it would be hastily discounted by the same people who fuss over the fact that he didn't. Why? Because just like most classical literature, Philo's works have been passed down to us by Christians.[150]

In the field of ancient history, arguments from silence are generally impotent, even in cases in which one has reason to believe an author should have written about a topic. Besides the fact that the bulk of ancient texts are lost, ancient writers did not take it upon themselves to write in the comprehensive manner that modern historians sometimes do.[151]

Numerous Confirmations of the Bible

We have already seen how ancient writings confirm some central elements of the Bible. Tacitus mentions Jesus along with particulars about when he lived, where he lived, that he was executed, and the name of the person who carried out the execution. Couple this with the evidence from Josephus, and it is certainly rational to conclude that Jesus walked the earth. Moreover, it is also plain that the Bible records a number of patently accurate details about Jesus' life.

This, however, is just a small portion of the available evidence. I haven't compiled an exhaustive list, but there are easily hundreds (maybe even thousands) of people, places, and events mentioned in the Bible that are corroborated by other ancient texts. A few of these involve major aspects of the Bible, but most are incidental details which, when taken together, make it clear that the Bible contains a wide range of verifiable facts. Consider the following:

- The Bible states Jesus was crucified in Jerusalem and that Pilate placed a message on his cross written in three languages: Hebrew, Greek, and Latin.[152] Several passages in Josephus reveal that these three languages were common in Jerusalem during this era.[153] [154] [155]

- The Bible states the Roman Emperor Claudius "commanded all Jews to depart from Rome."[156] [157] The writings of the Roman historian Suetonius are in agreement with this. He, too, wrote that Claudius "banished from Rome all the Jews."[158] [159]
- John the Baptist is a major figure in the New Testament. The Bible claims he was beheaded by a ruler named Herod.[160] The writings of Josephus corroborate the existence of both men and concur that Herod executed John.[161]
- The Bible speaks about events that transpired between Felix, the governor of Judea, and a Christian named Paul. Felix's wife, Drusilla, is also mentioned.[162] The existence of both Felix and Drusilla is corroborated by Tacitus.[163] [164]
- The Bible states that the Jewish sect referred to as the Sadducees did not believe there was an afterlife.[165] Josephus substantiates this:

> But the doctrine of the Sadducees is this: That souls die with the bodies….[166]

All this is not to say there aren't discrepancies between the Bible and other ancient writings. However, we can read stories in reputable publications about recent events and find pronounced discrepancies. Consider these headlines that appeared in *USA Today* and the *Washington Post* on February 24, 2003:[167]

Vaccine for AIDS Appears to Work.
– *USA Today,* first edition. Page A1.

Vaccine for AIDS Shows Promise.
– *USA Today,* final edition. Page A1.

1st AIDS Vaccine in Large Test Found to Be Mostly Ineffective.
– *Washington Post,* Page A2.

Likewise, here are two contradictory Yahoo News headlines for the very same Associated Press poll:[168]

AP poll: Americans optimistic for 2007.[169]

Poll: Americans see gloom, doom in 2007.[170]

When there are conflicts between the Bible and other ancient writings, some immediately conclude the Bible is inaccurate without even considering the prospect that the Bible is correct and the conflicting source is in error. Of course, the converse also applies. Nonetheless, despite the uncertainties inherent in the study of ancient history, the evidence is over-whelming that the Bible contains a wide range of factual content. Further-more, after years of study, I have yet to find one instance in which one can reasonably conclude the Bible is clearly inaccurate.

Exact Timing of the Crucifixion Confirmed by History and Astronomy

As we have already seen, the Bible contains chronological details, such as the names of prominent government officials and the times of their reigns, that correspond to chronological details in other ancient works. One of the most stringent tests of an ancient work's accuracy is using such details to calculate when certain events supposedly took place and then comparing the results with independent sources to see if they concur.

In this section, we will apply this test to the events surrounding the execu-tion of Jesus. Given the intricacies of ancient chronology, this type of inquiry is often fairly complicated, and what follows is no exception. Yet, understand-ing details is the often price we must pay for being informed instead of indoc-trinated. The reward for our effort will be genuine insight into how meticu-lously accurate various details of the Bible have proven to be.

Note that when one combines multiple facts, as we are about to do, multi-ple opportunities arise to make a mistake, but do not make the error of equat-ing complexity with a lack of certainty in the results. To quote the historian Isaac Taylor:

> The strength of evidence is not proportioned to its simplicity, or to the ease with which it may be apprehended by all persons; on the contrary, the most conclusive proof is often that which is the most intricate and complicated. … [T]he conclusion which perhaps not fifty men in Europe can, with full intelligence, know to be true, is actually as true as the axiom which the schoolboy comprehends at a glance.[171]

However, don't let these words create the impression that what follows is overly complex. Once again, just bear in mind that the sections of text enclosed by arrows (▶◀) and printed in a contrasting font are primarily for academic readers and can be skipped without missing crucial information or disrupting the flow of the book.

Let's begin with a list of claims in the Bible that bear upon the timing of the crucifixion:

1) Jesus was executed while Pontius Pilate was in office.[172]

2) Jesus was executed on a Friday.[173]

3) Jesus was Jewish,[174] and except for some time spent in Egypt as a child, the events of his life took place in the land of Israel.[175]

4) To celebrate the Jewish holy feast called Passover, lambs were slain in the first month of the year on the 14th day.[176] This month is called "Nisan," and it falls in the spring.[177] [178] Jewish writers of the first century also confirm they observed this practice.[179] [180]

> ▶In general, Jewish calendars reckon a 24-hour day from nightfall to nightfall,[181] [182] [183] as opposed to our calendar, which reckons a 24-hour day from midnight to midnight. Applying this convention to the Biblical command that the Passover lambs be slain on the 14th of Nisan "in the evening" (interpreted by the rabbis to mean from after noon until nightfall) and eaten "that night," the Passover feast took place on what we would consider to be the same day the lambs were slain in our calendar, but this was actually the next day in Jewish calendars, or the 15th of Nisan.[184] [185] [186] [187] [188] Therefore, when determining dates in ancient Jewish calendars, one must be careful with the word "Passover" and try to ascertain whether the writer was referring to the Passover feast or the slaying of the Passover lambs.◀

5) Jesus was crucified on the same day that the Passover lambs were slain or on the day afterwards. In other words, on the 14th or 15th day of the month Nisan. The Book of John states Jesus was crucified on the same day the Passover lambs were slain.[189] The Books of Matthew, Mark, and Luke state Jesus was crucified on the day after the Passover lambs were slain.[190] Contrary to what may seem to be the case at first glance, this does not show the Bible is contradictory and, therefore false. In fact, precisely the opposite is true. As we will see in Chapter 4, this is a consequence of two competing Jewish calendars that existed in this era.

6) Jesus was baptized no earlier than the "fifteenth year of the reign of Tiberius Caesar."[191]

7) There were at least three separate Passover celebrations from the time Jesus was baptized until the time he was crucified.[192]

Below is a list of assertions in other ancient works that also bear upon the timing of the crucifixion:

1) The Roman emperor Augustus died on August 19, 14 A.D.[193]

2) When Augustus passed on, Tiberius succeeded him as ruler of the Roman Empire.[194]

3) Pontius Pilate was removed from office and ordered to appear before Tiberius. As Pilate was on his way to Rome to appear before Tiberius, Tiberius died.[195]

4) Tiberius Caesar died on March 16, 37 A.D.[196]

5) A group called the Pharisees was the dominant Jewish sect during the time period in which the Bible and Tacitus place Jesus. The multitude of Jewish people followed the Pharisees' teachings.[197] Some of the historically significant Pharisees of this era were Josephus,[198] a rabbi by the name of Gamaliel, and his son, Simeon ben Gamaliel.[199] [200]

▶6) The calendar of the Pharisees was based upon astronomical observations and functioned in the following manner:

   a) The first day of each month started when the crescent of the new moon became visible.[201]

   About once every 29½ days, the moon becomes invisible to us on earth.[202] This occurs when the moon is lined up between the earth and the sun, as shown in the diagram below. (In this two-dimensional sketch, assume the earth, moon, and sun are not in the same plane. If they were, it would result in a solar eclipse.)

New Moon

Dark side of the moon faces the earth

© iStockphoto.com/Eraxion/janrysavy/knickohr

Since the dark side of the moon directly faces the earth, the moon is invisible to us. As the moon continues to orbit the earth, it moves out of the phase of being completely invisible and begins to appear as a thin crescent. It was the observance of this crescent that triggered the start of a new month. It takes approximately 354½ days for 12 of these months to elapse,[203] but it takes about 365¼ days for a solar year to elapse.[204] This means there is a 10¾-day discrepancy between 12 lunar months and one solar year, which leads to the next point.

b) An extra month was added in certain years to keep the calendar in step with the seasons.[205]

    This is similar to what we do with leap years with our modern calendar. A typical year is 365 days, but since it takes the earth 365¼ days to orbit the sun, every fourth year we add an extra day on February 29th. This keeps our calendar consistent with the seasons. If we did not add this extra day, over the course of 500 years, the month of June would be pushed into winter. Since a typical year in the calendar used by the Pharisees was 10¼ days shorter than a solar year, if they didn't add leap months, it would only take 12 years for the month they associated with the start of summer to be pushed into winter.

c) All months were either 29 or 30 days long.[206] [207]

d) The dividing line between days was nightfall or the "evening twilight."[208] This means each new day began around sunset, not at midnight. This is also the same time the new moon becomes visible.[209] Therefore, the new moon could have appeared during the final minutes of the waning day or during the opening minutes of the waxing day. Regardless, it was the waxing, not the waning day, that became the first day of the new month.[210] Practically speaking, if a new moon became visible on what we would consider to be a Tuesday evening in our modern calendar, the Pharisees would consider Wednesday to be the first day of the new month.

e) The 14th of Nisan (when the Passover lambs were slain) took place when the sun was in the constellation Aries.[211]

f) Two days later, on the 16th of Nisan, the Pharisees performed a religious ceremony that required the use of fresh barley. It was forbidden to harvest the spring crops until this ceremony was performed.[212] [213] ◄

CORRELATING THE EVIDENCE

The assertions above constrain one another and thus delimit timeframes for the crucifixion. Note, we will not assume any of the Biblical or other claims are accurate. For now, the only objective is to determine their implications.

Let's start with the broad limits and progressively work toward more narrow timeframes. The first step is to find an early boundary. Four books in the Bible focus upon the earthly life of Jesus. These are the Books of Matthew, Mark, Luke, and John—otherwise known as the four Gospels. The third chapter in the Book of Luke begins with the following chronological detail:

Now in the fifteenth year of the reign of Tiberius Caesar … the word
of God came unto John the son of Zacharias in the wilderness.[214]

This refers to the start of the ministry of John the Baptist, who preached
in the wilderness of Judea and baptized many people, one of whom was
Jesus.[215] Hence, the Bible provides an early limit for Jesus' baptism: "the
fifteenth year of the reign of Tiberius Caesar."

Next, we must identify this timeframe. Tiberius was a Roman emperor
and a major figure in history. His predecessor (Augustus Caesar) died
on August 19, 14 A.D.,[216] which is therefore the earliest date Tiberius
could have begun to reign. ▶Some have made the claim that Tiberius
and Augustus co-reigned for several years before the death of Augustus.
However, ancient historical works, coins, and inscriptions prove beyond a
reasonable doubt that the reign of Tiberius was reckoned from the death of
Augustus or shortly afterwards.[217 218 219]◀

Even though we have a precise date to start with, there are different
ways of interpreting what "the fifteenth year" might mean. ▶The most
obvious is to define the first year and all successive years as beginning on
the exact date that Augustus died. Thus, the first year of Tiberius would have
been from 8/19/14 to 8/18/15, the second year from 8/19/15 to 8/18/16,
and the "the fifteenth year" from 8/19/28 to 8/18/29. Some ancient
authors used this reckoning, and the result fits perfectly with the rest of
the evidence we are going to examine. Some authors, however, used other
methods, including starting the first year on the day Tiberius began to reign
and starting the second year and all successive years on New Year's Day. An
added complication is that different cultures celebrated the New Year on
different days.◀ Since we are trying to establish an early limit, we'll use the
earliest possible reckoning. This means "the fifteenth year of the reign of
Tiberius Caesar" began no earlier than the fall of 27 A.D.[220 221 222]

The Gospels mention three separate Passovers between the baptism
and crucifixion of Jesus,[223] effectively asserting there were at least three
Passovers during this period.[224] Passover took place in the spring,[225 226 227
228 229] and if we move ahead three springtimes from the fall of 27, we arrive
at the spring of 30 A.D. Thus, if we push the chronological markers in the
Bible to their earliest possible limits, we conclude it places the crucifixion
of Jesus no earlier than the spring of 30 A.D.

Calculating the later limit is easier. The Bible and Tacitus assert Jesus was
executed under Pontius Pilate.[230] Josephus wrote that Pilate was removed
from office and ordered to appear before the emperor Tiberius. He also
wrote that as Pilate was on his way to Rome to appear before Tiberius,
Tiberius died.[231] Thus, the death of Tiberius marks the latest possible limit

for the execution of Jesus. Tiberius died on March 16, 37 A.D.[232]

Bringing all of the above together, the time markers in the Bible and other ancient works lead us to believe Jesus was crucified sometime between the spring of 30 A.D. and March 16, 37 A.D.

The Bible claims the crucifixion took place on the same day the Passover lambs were slain or the day afterwards.[233] [234] In the Jewish calendar, this is the same as saying it happened on 14th or 15th day of the month of Nisan.[235] [236] [237] [238] These are exact dates, just like any dates in our calendar. The Bible also claims the crucifixion took place on a Friday.[239] These are key assertions because they allow us to get very specific.

During the era in which the Bible, Josephus, and Tacitus place Jesus, the vast majority of Jews followed a sect known as the Pharisees.[240] Information about the calendar they used has been preserved in several ancient works. Since this calendar was based upon astronomical observations, it can be reconstructed using the science of astronomy. Moreover, we can determine on which day of the week any date in this calendar fell because the seven-day cycle of weeks we use today has continued uninterrupted for at least 3,000 years.[241] [242] Continuing in our correlation of the evidence detailed above, over the next several pages of academic text, we will employ historical and astronomical knowledge to reconstruct this calendar.

▶In the calendar of the Pharisees, the first day of each month took place when the crescent of the new moon became visible.[243] Hence, our first task is to identify every new moon that could have triggered the start of the month Nisan in the period from 30 to 37 A.D.

The Bible and numerous Jewish sources from ancient up through modern times unanimously associate the month of Nisan with the spring.[244] [245] [246] [247] [248] To get more specific, consider the following: It takes about 354½ days for twelve lunar months to elapse, while it takes about 365¼ days for a solar year to elapse.[249] [250] [251] So, in order to keep their calendar in step with the seasons, the Pharisees added an extra month in certain years.[252] If this weren't done, Nisan would naturally drift towards the winter season. This would have created a major problem because on the 16th of Nisan, the Pharisees performed a religious ceremony that required the use of fresh barley. If this ceremony occurred in winter, no barley would be available to perform the ceremony. Furthermore, it was forbidden to harvest the spring crops until this ceremony was performed.[253] [254] Therefore, leap months could not be carelessly added to the calendar because this would shift Nisan toward the summer and the crops would become overripe before harvesting was permitted. In sum, the requirements associated with this ceremony created a situation wherein the month of Nisan had to fall within certain boundaries of the solar year.

In his autobiography, Josephus writes that he conducted himself "according to the rules" of the Pharisees. Information he records allows us to pin down the month of Nisan:

> In the month of Xanthicus, which is by us called Nisan, and is the beginning of our year, on the fourteenth day of the lunar month, when the sun is in Aries, (for in this month it was that we were delivered from bondage under the Egyptians,) the law ordained that we should every year slay that sacrifice which I before told you we slew when we came out of Egypt, and which was called the Passover....[255]

What this tells us is that the 14th of Nisan occurred when the sun was in the constellation Aries. With astronomy software, we can easily determine when this was.[256] Using the Julian calendar as our frame of reference,[257] the sun was in the constellation Aries from the 24th of March through the 18th of April during the years 30–37 AD.[258] Although unneeded for our solution, we will include a seven-day margin of error on both sides of these dates to make sure we have examined the broadest realistic range of dates. This results in a range from March 17th to April 25th.[259]

The next step is to determine which new moons fit the criterion above. By definition, in the calendar of the Pharisees, the sighting of the new moon was the first day of the month. Since the information above tells us when the 14th of Nisan occurred, we can apply it to the first of Nisan simply by subtracting 13 days. When this is done, we find that during the years 30–37 A.D., the first of Nisan occurred no earlier than March 4th and no later than April 12th. We can now proceed to determine exactly when new moons appeared during this timeframe.

The determination of when new moons become visible has been calculated with relatively good accuracy since 1910.[260] Today, we have access to software that calculates the appearance of new moons with fantastic precision.[261] It allows us to vary such parameters as the location from which the moon is viewed and climate conditions such as temperature and humidity. The program does not account for cloudy weather that may have totally obstructed the new moon, but we will do this for ourselves. In the meantime, by using two sets of conditions—one that is favorable for seeing the new moon and one that is unfavorable—we can establish a list of viable dates for the first of Nisan.[262]

Using this software, we obtain 16 dates on which the new moon could have been visible between March 4th and April 12th in the years 30–37.[263] These are Julian dates, and as such, they correspond to a 24-hour day that lasts from midnight to midnight. However, in the calendar used by the Pharisees, the dividing line between days was the "evening twilight."[264]

This is also the same time that the new moon becomes visible.[265] When the new moon was sighted, the day that had just begun or was about to begin became the first day of the month.[266] Therefore, we must index each of these dates by one day in order to bring them into line with the calendar used by the Pharisees.[267]

The fact that cloudy weather may have obstructed the new moon makes it possible that the days following those calculated above are also viable dates for the first of Nisan. This was unlikely, however, given that around 100 A.D., Tacitus wrote that rain was "uncommon" in Judea.[268] [269] Still, it is a possibility, and we will account for it. Also, realize that in the calendar of the Pharisees, all months had to be either 29 or 30 days long.[270] [271] Thus, even consecutive days of cloudy weather could not delay the start of a new month by more than one day because this would cause the outgoing month to be more than 30 days long, which was forbidden. Accounting for this scenario, there are 27 potential dates for 1st of Nisan in the years 30–37.[272] With this information, we can determine all viable dates for the 14th and 15th of Nisan, which are the two dates the Bible provides for the crucifixion of Jesus.

Year	All Viable Dates for the 1st of Nisan	All Viable Dates for the 14th of Nisan (Book of John)	All Viable Dates for the 15th of Nisan (Books of Matthew, Mark, and Luke)
30	Fri Mar 24, Sat Mar 25, Sun Mar 26	Thu Apr 6, Fri Apr 7, Sat Apr 8	Fri Apr 7, Sat Apr 8, Sun Apr 9
31	Wed Mar 14, Thu Mar 15, Thu Apr 12, Fri Apr 13, Sat Apr 14	Tues Mar 27, Wed Mar 28, Wed Apr 25, Thu Apr 26, Fri Apr 27	Wed Mar 28, Thu Mar 29, Thu Apr 26, Fri Apr 27, Sat Apr 28
32	Mon Mar 31, Tues Apr 1, Wed Apr 2	Sun Apr 13, Mon Apr 14, Tues Apr 15	Mon Apr 14, Tues Apr 15, Wed Apr 16
33	Sat Mar 21, Sun Mar 22	Fri Apr 3, Sat Apr 4	Sat Apr 4, Sun Apr 5
34	Thu Mar 11, Fri Mar 12, Fri Apr 9, Sat Apr 10	Wed Mar 24, Thu Mar 25, Thu Apr 22, Fri Apr 23	Thu Mar 25, Fri Mar 26, Fri Apr 23, Sat Apr 24
35	Wed Mar 30, Thu Mar 31	Tues Apr 12, Wed Apr 13	Wed Apr 13, Thu Apr 14
36	Sun Mar 18, Mon Mar 19, Tues Mar 20	Sat Mar 31, Sun Apr 1, Mon Apr 2	Sun Apr 1, Mon Apr 2, Tues Apr 3
37	Thu Mar 7, Fri Mar 8, Sat Mar 9, Sat Apr 6, Sun Apr 7	Wed Mar 20, Thu Mar 21, Fri Mar 22, Tues Apr 9, Wed Apr 10	Thu Mar 21, Fri Mar 22, Sat Mar 23, Wed Apr 10, Thu Apr 11

◄

The fact that the Bible places the crucifixion on a Friday narrows this list down significantly. Furthermore, we can discard all dates from the year

37 because they fall after the latest possible day Pontius Pilate was removed from office (March 16, 37). This leaves us with only five plausible dates:

April 7, 30
April 27, 31
April 3, 33
March 26, 34
April 23, 34

While bearing in mind that we are testing the Bible for chronological accuracy, realize the dates above were arrived at by reconstructing the calendar of the Pharisees in order to determine every possible date for the 14th and 15th of Nisan that fell on a Friday between the years 30 and 37 A.D. If the Bible and the historical citations above are accurate, these are the only dates on which Jesus could have been executed. Given that we went to such great lengths to include all reasonable possibilities in our calculations, these are very specific results. It is also worth noting that we would have only arrived at two of them were it not for the various factors of safety that were incorporated:

April 7, 30
April 3, 33[273]

As it turns out, these dates have been identified for quite some time as possibilities for the crucifixion. In 1910, a person by the name of J.K. Fotheringham used the most recent findings of astronomy to pin down these dates with a good deal of precision.[274] Since then, astronomical knowledge has increased, allowing for greater and greater accuracy, and various scholars have taken a stab at the calculations using different historical and astronomical methods.[275][276][277][278] Still, the results have almost always boiled down to these two dates.

So, why did we bother to go through all this? Although others have done exceptional work in this area—and I have used this work as a basic framework for my calculations—I did not feel comfortable with some of the methodologies that were employed. There is no need to go through a list of the details, but suffice it to say I wanted every historical assertion thoroughly documented and every plausible scenario accounted for. (Serendipitously, this process of research uncovered information that we will use in Chapter 4 to unravel a mystery that has been pondered for at least 1,800 years.)

Under a Blood-Red Moon

Here is the astonishing part. In 1983, a professor and research assistant at the Department of Metallurgy and Science of Materials at Oxford University published an article in *Nature* magazine that arrived at three "calendrically possible dates" for the crucifixion.[279] They are all among the dates we calculated:

April 7, 30
April 3, 33
April 23, 34

Even though these men employed more precise astronomical data than had been used previously, they, like others, ended up with the same dates. However, they made an astute observation about another passage in the Bible describing an event set 50 days after the crucifixion.[280] The Bible claims that while addressing a hostile audience in Jerusalem, a disciple of Jesus named Peter quoted a prophecy from the Old Testament book of Joel, a work written hundreds of years before Jesus was born.[281] Among other things, this book predicted, "The sun will be turned to darkness and the moon to blood before the coming of the great and dreadful day of the LORD."[282] After citing this passage, Peter told the crowd they had witnessed these events.[283]

The sun turning to darkness is an obvious reference to the crucifixion, as the Bible claims in three places that the sun was darkened while Jesus was on the cross.[284] However, there is no readily apparent explanation for what is meant by the moon turning "to blood." Common sense would tell us this is a myth or some sort of strange imagery. In either case, it is the type of thing most people read over and quickly disregard.

The Oxford researchers, however, recognized that the phrase "moon to blood" may refer to a lunar eclipse. This is because the moon actually turns into a dark red color during certain lunar eclipses. ▶Why? Though sunlight appears to be white, it is comprised of all the colors of a rainbow.

Visible Light Spectrum

Shorter Wavelegth   **Longer Wavelength**

As sunlight passes through the earth's atmosphere, it bounces off air molecules. This causes the light to scatter. Note from the diagram above that blue light has a shorter wavelength than red light. Thus, when sunlight travels through the earth's atmosphere, the blue light is more likely to bounce off air molecules and scatter. Conversely, the red light is less likely to scatter and continues on a straight path. This is why the sky is blue and why the sun looks red as it sets.[285] This phenomenon can have a striking effect on the moon's color during lunar eclipses.◄

Here is a picture of this dramatic phenomenon during a 2003 eclipse:

© iStockphoto.com/knickohr

The Oxford researchers also found three examples from other ancient historical texts in which similar verbiage was used in reference to a lunar eclipse:

[The moon was] suffused with the color of blood. (331 B.C.)

The Moon was turned to blood. (304 A.D.)

The Full Moon was turned to blood. (462 A.D.)

Based upon this evidence, they checked to see if such an eclipse occurred on any of the three plausible dates for the crucifixion. Sure enough, on Friday, April 3, 33 A.D., there was a partial lunar eclipse visible over Jerusalem during which the eclipsed area was red and the rest of the moon "yellow-orange." Astronomy software allows us to render the sky in any place and at any time in earth's history, and as such, a close-up rendering of the moon during this very event is shown here using software called Starry Night:

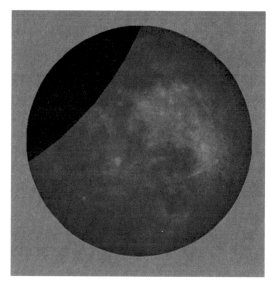

April 3, 33 A.D., 6:00 PM

For contrast, this is how Starry Night renders the moon when conditions are similar and there is no lunar eclipse:

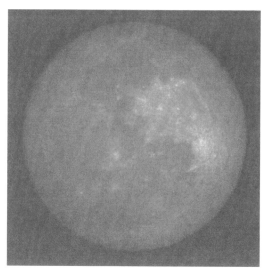

March 6, 41 A.D., 5:45 PM

This is nothing short of remarkable. What we have here is strong evidence that:

1) The Bible is describing an actual event when it asserts the moon turned to "blood."

2) Numerous independent chronological details in the Bible and in other historical sources are accurate down to the very day.

It should also be noted that Friday, April 3, 33 A.D. falls right in the middle of the evidence we examined.[286] If we didn't push this evidence to its limits, allow for possibilities such as cloudy weather, or use safety factors, we would have arrived at this date and only one other.

In sum, using numerous Biblical and historical assertions to calculate every plausible timeframe for the crucifixion yields only five specific dates. The Bible claims that 50 days after the crucifixion, the apostle Peter told a crowd that the moon turned to blood and that they witnessed it. On one of these five dates, specifically the one that sits in the center of the evidence, this claim is confirmed in striking detail by an astronomical event that accords with a Biblical prophecy written hundreds of years beforehand.

ONLY ROMANS COULD ADMINISTER THE DEATH PENALTY

Now that we have established a date for the crucifixion, we are in a position to learn even more. An ancient Jewish work called the Jerusalem Talmud states the following:

> It was taught: Forty years before the destruction of the Temple the right to judge capital cases was withdrawn….[287]

During the first century A.D., the land of Israel was under Roman rule. However, the Romans allowed the Jewish people to govern certain aspects of their country.[288] The Jerusalem Talmud states that 40 years before the Temple was destroyed, the right to judge cases involving the death penalty was withdrawn. According to Josephus, the Temple was destroyed in 70 A.D.[289] This means this policy became effective in about 30 A.D., just a few years before Jesus was crucified. In the Gospel of John, we read that when Jewish officials took Jesus to Pontius Pilate, the following exchange occurred:

> Then Pilate said to them, "You take Him and judge Him according to your law." Therefore the Jews said to him, "It is not lawful for us to put anyone to death…."[290]

This passage fits perfectly with the Jerusalem Talmud and is further compelling evidence that the Bible recounts real events and provides accurate details about them.

There is a great deal more we could explore from here, but the constraints of space and time oblige us to move on. Thus far, we have primarily examined the New Testament, which is the portion of the Bible that is uniquely Christian. In the next two chapters, we will delve into the Old Testament, which both Christians and Jews consider sacred.[291]

CHAPTER 2

SCIENCE

Even before entering high school, by using the expansive knowledge and flawless rationale many a 13-year-old believes he possesses, I arrived at the firm conclusion there was no such thing as God. This was a view I maintained through college and up until my mid-twenties. Given that I considered the Bible a book of myths and fairy tales, it mystified me how any reasonable and intelligent person could treat it as a source of knowledge. Yet, I had some friends who were smart, level-headed, and also happened to be Christians. One day, while traveling with two such friends (husband and wife), one of them asked me to get something out of the glove box of their car. While doing so, I noticed a Bible and asked if I could take a look at it. Randomly opening to the book called Proverbs, I found a variety of insightful sayings about human nature. If memory serves me correctly, one example of what I read was:

> Though you grind a fool in a mortar, grinding him like grain with a pestle, you will not remove his folly from him.[292]

I remarked upon the amusing nature of these sayings, and my friends offered me their Bible. I refused several times but they persisted, and so I took it home. Over the course of the next year, I read that Bible from cover to cover and was intrigued by what I found. As a person who has been criticized for putting too much emphasis on facts and data while disregarding the instinctive side of human nature, I expected these ancient writings to be bursting with numerous false statements about the physical laws that govern our universe. Instead, what I found was uncannily consistent with science. For instance, while in college, I took a course entitled "The Ocean" in which we learned about the nature of waves. Before reading further, stop and ask yourself, "What causes the waves you see at the beach?"

On one occasion, I called ten people with an assortment of bachelor's degrees and posed that question to them. Two of them said they had no idea, and eight said waves are caused by the moon. This is a fairly common misconception, but in fact, the gravitational pull of the moon causes tides, not waves.[293] (Technically speaking, tides are a waveform, and under certain special conditions they can contribute to some very sizeable waves,[294] but this is clearly not what is meant by "the waves you see at the beach.")

In "The Ocean," I learned waves at the beach are primarily caused by wind.[295] Yet, the belief that the moon causes them is so engrained that someone once spent ten minutes insisting to me this was the case. Thus, if you are one who adamantly believes this, consider the following from the *Encyclopædia Britannica*:

> Waves on the sea surface are generated by the action of the wind.[296]

Even living during this supposedly enlightened age in which we have PBS, *National Geographic*, the Internet, *Discovery Channel*, and access to 13 years of public education now costing about $140,000 per student in the U.S.,[297] most people don't answer this question correctly. Yet, the Bible had it right thousands of years ago. As the Book of Psalms states:

> For He commands and raises the stormy wind, which lifts up the waves of the sea.[298]

Over the following pages, we will examine more information such as this. Realizing it is impossible to cover all aspects of how the Bible relates to science in a single book let alone one chapter, we will explore a diverse sampling of topics, including earth science, Galileo, physiology, life after death, miracles, and neurobiology. After this, we will consider remarkable insights about the existence of God from some of the most accomplished scientists in history and some gifted scientists of today.

What Holds Up the Earth?

Today, we all know the earth is not supported by any physical structure. Astonishingly, the following passage in the Old Testament book of Job spelled this out millennia ago:

> He stretches out the north over empty space; He hangs the earth on nothing.[299]

When I was an atheist, I read the entire Bible keeping an eye out for scientific flaws and failed to find them. Since becoming a Christian, I've continued to look for such errors and have yet to discover one. Note that in several instances I thought I did at first glance, but further research proved quite the opposite. For instance, the Bible states:

> For the pillars of the earth are the LORD's, And He has set the world upon them.[300]

Does this passage conflict with the one quoted above and assert the earth is supported in space by marble columns in a manner perhaps such as this?

© iStockphoto.com/janrysavy/frederickdesign

Far from it. The critical point to realize is that this portion of the Bible was written in Hebrew, and if we analyze the word "pillars," which is translated from the Hebrew word "matsuwq," we find its meaning is uncertain.[301] Depending upon which Bible translation one reads, the word can be translated "pillars," "foundations," or "fixtures."[302] One scholar even translates it as "straits."[303]

By examining the context, various scholars have attempted to arrive at an accurate translation, but this is no easy task because there are two known roots from which this word may be derived, and neither fits well with the context. One of these roots literally means "to pour out." Using this definition and bearing in mind that whatever this Hebrew word means, the context states it is something God "has set the world upon," some translators think the word must refer to a molten material that was poured into a mold to create supports.[304] Thus, they arrive at words such as "pillars," "foundations," and "fixtures." The other potential root means "narrow," and based upon this, one scholar takes this to be a reference to underground rivers referred to as "narrows" or "straits."[305]

Obviously, none of these explanations is very convincing, which highlights the fact that we are only working with best guesses. Yet, there is another very important point to consider. Although it is not obvious to someone reading an English translation of the Bible, the sentence we are examining and the surrounding text are the lyrics to a song.[306] Given all this, it would be ridiculous to claim the Bible asserts the earth sits on pillars like a golf ball sits on top of a tee. Maybe 3,000 years in the future, if someone digs up a recording of the song "We Will Rock You," he or she will conclude our society made a habit of stoning people.

GALILEO

In all seriousness, we should be very careful when relating the Bible to science, and key to doing this is being mindful of the context. Obviously, reading adjacent text is critical to understanding the context of any writing, but when it comes to writings from ancient or foreign cultures, we must often go further than this. For example, as we'll discuss in Chapter 10, understanding certain aspects of the civilizations in which a writing is set is also important. Of course, we should keep the general context in mind as well. One can read any portion of the Bible and plainly see its main purpose is not to teach science. Thus, to avoid the mistake of misinterpreting the Bible, we must recognize the fact that it is not a physics textbook.

This issue was at the center of an argument that occurred in the 17th century between Galileo and the Roman Catholic Church. Galileo's research led him to conclude that the earth moved around the sun, but the Catholic Church took the opposite view, and quotes from the Bible such as the following were used to support the claim that the sun moved around the earth:[307]

> [The sun] is like a bridegroom coming out of his chamber, And rejoices like a strong man to run its race. Its rising is from one end of heaven, And its circuit to the other end; And there is nothing hidden from its heat.[308] [309]

One could choose to interpret this passage as a technical explanation of the sun's movements—or as a poetic description of what we see when we look at the sun. The rational choice is very apparent. Likewise, when we in modern times say the sun rises or sets, we are not claiming that the sun orbits around the earth. And when we use the word "moonlight," we are not stating or implying that the moon generates light. So why would anyone interpret this passage in the illogical manner they did?

To answer this question, we need to delve inside the minds of others, which always involves some guesswork, but the historical record provides ample evidence with which to do so.

In the era of Galileo, the writings of the ancient Greek philosopher Aristotle formed the primary basis of science education in Europe.[310] The premier educators were a Catholic group called the Jesuits, and their science curriculum was built upon the teachings of Aristotle, who claimed that the earth was stationary and the rest of the universe moved around it.[311] [312] [313] Instead of objectively assessing Galileo's theory, the Jesuits saw his conflicting view as a threat to their educational system and convinced the leadership of the Catholic Church to take a stance against it.[314] [315] [316] Given that the Jesuits were wed to Aristotle's position, we can only surmise they were not reading the Bible to grasp what it said but to justify that which they already believed.

In 1615, Galileo wrote a letter to the Grand Duchess of Tuscan, Italy, in which he quoted from a prominent Christian of antiquity known as Saint Augustine. Take an extra moment to thoroughly grasp the forthcoming extract because it incisively rounds out this issue:

> In points that are obscure, or far from clear, if we should read anything in the Bible that may allow of several constructions consistently with the faith to be taught, let us not commit ourselves to any one of these with such precipitous obstinacy that when, perhaps, the truth is more diligently searched into, this may fall to the ground, and we with it. Then we would indeed be seen to have contended not for the sense of divine Scripture, but for our own ideas by wanting something of ours to be the sense of Scripture when we should rather want the meaning of Scripture to be ours.[317]

PHYSIOLOGY

Although the human body is incredibly complex, the Bible reveals insights about its functioning that science has only recently confirmed. Among these, the Book of Proverbs states, "A merry heart does good, like medicine…."[318]

A medical study published in 2003 convincingly supports this statement. A team of trained interviewers evaluated the emotional state of 354 physically healthy people. This was done by interviewing each person on seven different days. In these interviews, people were asked to rate on a scale of 0 to 4 the level at which they experienced the following emotions that day: lively, full-of-pep, energetic, happy, pleased, cheerful, at ease, calm, relaxed. The first six of these feelings are clearly consistent with a "merry heart," while the last three may or may not fall into this category. Based upon their responses, each person was given a score.

Next, the people were independently quarantined for a day, given nose drops containing common cold viruses, and then quarantined for five more days. Every day they were tested to see if they had a cold. The result was that people with high positive emotion scores were about 60% less likely to develop a cold than the people with low scores. After controlling for factors such as presence of antibodies in the bloodstream that provide resistance to the type of viruses administered, age, sex, and body mass, researchers found that people with high positive emotion scores were 2.9 times less likely to catch a cold than people with low scores.[319] Those unfamiliar with medical studies should take note that this degree of efficacy is remarkable. For comparison, here is how the Mayo Clinic summarizes the results of more than 30 clinical studies examining the effects" of taking Vitamin C as a cold preventative:

> Overall, no significant reduction in the risk of developing colds has been observed. … Notably, a subset of studies in people living in extreme circumstances, including soldiers in sub-arctic exercises, skiers, and marathon runners, have reported a significant reduction in the risk of developing a cold of approximately 50%.[320]

Yet, there is more. The rest of the sentence from Proverbs quoted above reads:

> But a broken spirit dries the bones.[321] {From the context provided by the first half of this sentence, we can see that "dries the bones" is a metaphor referring to ill health or weakness.[322]}

It would be intuitive to think the study we just reviewed also supports this statement, but it does not. It found that positive emotion scores were somewhat independent of negative emotion scores. In other words, just because someone has a lot of positive emotions doesn't necessarily mean he or she has few negative ones. This study found no correlation between negative emotions and the rate at which people contracted a cold. However, another study published in the *New England Journal of Medicine* found that "[p]sychological stress was associated in a dose-response manner with an increased risk of acute infectious respiratory illness."[323]

Furthermore, a study in the *Journal of Experimental Medicine* identified a "sophisticated molecular mechanism regulating immune cell functions that can lead to stress-induced immunosuppression."[324] Thus, modern science provides ample evidence that "a broken spirit dries the bones."

Life after Death

The Bible asserts that we each have a soul that continues to exist after our body gives way.[325] Due to the advances of modern medicine, more and more people are being revived after their hearts have stopped beating and after an instrument called an electroencephalograph can no longer measure electrical activity in their brains. After resuscitation, many people describe similar vivid experiences, such as seeing departed family and friends, traveling through a tunnel, and being in the presence of a light that spoke with them.[326] These are referred to as "near-death experiences." Some suggest they are simply hallucinations induced by a lack of oxygen to the brain or by the drugs often given to patients in critical condition.[327] I don't doubt this may be the case in some instances, but certain episodes of this nature are not so easily explained away.

Before we go further, it should be pointed out that the Bible strongly condemns mystical and occult practices such as astrology, conjuring spirits, and witchcraft.[328] I mention this because such practices are sometimes associated with near-death experiences, and the clear difference between them should be recognized.

A paper published in *The Lancet*, a prominent British medical journal, documents an event that took place at one of ten Dutch hospitals involved with a study of near-death experiences. A 44-year-old man was brought into the hospital via ambulance. He was in a coma, and his skin had turned blue. He was given artificial respiration, heart massage, and defibrillation (electric shocks to restart the heart). After an hour and a half of treatment, the man's heartbeat and blood pressure returned to safe levels and he was transferred to another room, although he was still in a coma and hooked to a respirator.

The patient eventually became conscious, and a week later, one of the nurses who had cared for him while he was in critical condition saw him for the first time since then. The patient immediately recognized the nurse and asserted that the nurse would know where his dentures were. Since the nurse had removed the patient's dentures while he was unconscious, the nurse asked how he knew this. The man said he had seen the events of his treatment from a vantage point above where the doctors and nurses were.

He correctly stated the nurse put his dentures into a sliding draw under a cart with many bottles on it. He also accurately described the room, the people in it, and portions of the conversation that took place.[329]

Without going through a litany of such examples, suffice it to say that in assessing the scientific literature on near-death experiences, I would not characterize this body of research as ironclad proof of life after death but rather as credible evidence that makes it a very real possibility. Along similar lines, one month before he passed on, Albert Einstein expressed his condolences to the family of a deceased colleague in a letter that he concluded with these words:

> For those of us who believe in physics, this separation between past, present, and future is only an illusion, albeit a stubborn one.[330]

Although Einstein brought mankind's understanding of time to a new zenith with his special and general theories of relativity, his words here reveal that he considered the fundamental concept of time to be an illusion that lay outside the grasp of human intellect. For Einstein, this was truly about physics, not spirituality, for he rejected both God and the Bible.[331] Yet, the implications of his view about the illusory nature of time are manifest. Could it be there is a timeless reality that exists alongside the dimensions of space and time that our physical bodies inhabit? Let's examine further evidence on this subject.

Miracles

The strongest objection to the scientific accuracy of the Bible has to do with the many miracles it speaks of. Among these, the Bible states God divided the waters of the Red Sea and Jesus walked on water and rose from the dead.[332] Is it possible such events took place? One can argue miracles and science are incompatible simply because, by definition, miracles are events that contradict a known law of science.[333] However, to dismiss the possibility of miracles is circular reasoning. Such a mindset is based on the assumption that there is no God, and therefore miracles cannot happen, and therefore there is no God.

To put it another way, if there is a God who made the universe, He can certainly alter any aspect of it when and as He pleases. The fact that miracles don't take place before our eyes on a regular basis does not mean they never have and never do. In the words of a legal scholar who will be probed in an upcoming chapter:

While unbounded credulity is the attribute of weak minds, which seldom think or reason at all … unlimited skepticism belongs only to those, who make their own knowledge and observation the exclusive standard of probability. Thus the king of Siam [former name of Thailand] rejected the testimony of the Dutch ambassador, that, in his country, water sometimes congealed into a solid mass; for it was utterly contrary to his own experience.[334]

In reality, many well-documented occurrences can best be described as miraculous. One example is of a woman named Kelly Barker. In September 2003, Kelly was struck in the head by the side-view mirror of a moving pickup truck. The resultant brain damage was extremely severe, and her family sought the opinions of seven doctors, all of whom determined she would be in a vegetative state the rest of her life.

Discussion ensued over whether or not she should have her feeding tube removed, which would result in death by dehydration. Yet, her mother adamantly declared, "[T]here are so many people praying for Kelly that it wouldn't be right to not wait for some kind of miracle." About three months later, Kelly abruptly moved her legs off her bed and tried to stand up. A few months after this, she was talking and walking. In the words of two of her doctors:

> We have no real scientific explanation for Kelly's recovery, which is why I can say it is pretty miraculous. … [T]here wasn't even a question of recovery. The question then was: Do we maintain life or do we let nature take its course?
> – *Aaron Ellenbogen* (Neurologist)

> If 100 physicians examined Kelly after the accident, they would tell you she was going to be brain-dead for the rest of her life.
> – *Paul Jackson* (Family Physician)[335] [336]

We could easily fill this entire book with cases like this, and some people would just dismiss them all on the basis of happenstance, but the accumulated weight of such events is difficult to ignore. In fact, a recent nationwide survey of 1,100 physicians in the United States found 73% think miracles take place. Moreover, 55% claim to have personally witnessed such events in the lives of their patients. Even more surprising than this is the fact that only 12% of doctors think the miracles in the Bible are not true whereas 50% think they are metaphorically true and

37% think they are literally true.[337]

NEUROSCIENCE[338]

With regard to Biblical miracles, evidence from the science of neuro-biology corroborates the following narrative in the Book of Mark:

> [T]hey brought a blind man to [Jesus] and begged Him to touch him. So He took the blind man by the hand and led him out of the town. And when He had spit on his eyes and put His hands on him, He asked him if he saw anything. And he looked up and said, "I see men like trees, walking.[339]

The statement, "I see men like trees, walking," doesn't seem to make much sense, but as we will see, it is a crucial detail providing a scientific foothold that can be used to weigh the veracity of this account.

When eye surgeries restore sight to people who have been blind for a long time, some bizarre things happen. The reason is that it takes time for the brain to learn how to make sense of the images sent to it by the eyes.[340] Let's examine the separate cases of two men who lost their vision as children and had it restored more than 40 years later. In their first moments of sight, neither realized they were staring at a human face. But since they knew that voices come from faces, when someone spoke, each concluded the object he was looking at must be a face.[341] A few days later, one of them remarked that "trees didn't look like anything on earth."[342] Five weeks afterwards, this person had trouble distinguishing his dog from his cat,[343] [344] even though the vision in one of his eyes was 20/100.[345] One year after the other man's surgery, he could not identify facial expressions or tell apart one person from another, even though his eyesight was totally adequate to do so.[346]

The obvious point is that the statement of the man in the Bible remarkably parallels the experiences of people who undergo modern-day medical procedures that heal their vision. The eyes see clearly, but the brain doesn't, and the person tries to sort through the confusion by interpretation. This condition is common in people who acquire sight after extended periods of blindness.[347] Even when it seems the brain has adjusted, much of this adjustment takes the form of learning how to deduce what's going on instead of truly observing it.[348] Also noteworthy is the fact that moving objects are especially confusing to people who are new to seeing.[349] Consider what must go through someone's mind in order to say he sees men like trees, walking: "I know I am looking at people because I can hear their voices and see them moving, but they look like trees to me." This is precisely the type of phenomenon that takes place when someone's vision is healed.

What the Bible claims happened next is also significant. After the man spoke, "Jesus put His hands on his eyes again and made him look up. And he was restored and saw everyone clearly."[350] Without modern medical knowledge of what actually occurs when a blind person's vision is healed, the only logical way to interpret this passage is to deduce that Jesus did a poor job of healing the man's eyes on the first shot and had to try again. This is hardly the kind of story someone would make up if they wanted to portray Jesus as the "Son of God."[351] However, the science of neurobiology provides a context in which these events make sense. With the first touch, Jesus healed the man's eyes and with the second, enabled his brain to discern visual stimuli. If someone were all-knowing, capable of miracles, and desiring to leave evidence of this for future generations, are these not the type of clues He might leave for us?

Given what science has shown us about eyesight as it relates to brain function, this whole passage displays characteristics consistent with a real event that was accurately reported. Because of the miraculous nature of this account, it would be prudent to be skeptical, but given the scientific substantiation weighing in favor of its authenticity coupled with the abundant evidences supporting the Bible's truthfulness, it would also be rational not to randomly set it aside.

The Wonder of It All[352]

In my view, the most compelling evidence for miracles is the very existence of our universe and the marvels found within it. We will address this subject in great depth in Chapters 6 through 9. Although the information we will examine clearly falls within the scope of the current chapter, the vast amount of scientific data covered warrants several chapters. In the meantime, however, we shall bridge the subject with the words of some of the most accomplished scientists in history. It's not that these scientists or any others were infallible, but for those under the impression that science and Christianity are incompatible, their views are worth examining.

Undeniably, one of the greatest and arguably the greatest scientist of all time is Isaac Newton.[353] [354] [355] Newton is commonly known for mathematically defining gravity, but he is also one of two people who independently created the discipline of calculus. Most importantly, he developed three laws of motion that have been described as "the greatest single advance in human knowledge ever made."[356] [357] The *New Millennium Encyclopedia* states Newton "established the principal outlines of the system of natural science that has since dominated Western thought."[358]

Even though I have a degree in a scientific discipline that is highly dependent upon Newton's laws, only recently did I learn that almost half of Newton's writings were about the Bible and that he wrote statements such as these:[359]

> Did blind chance know that there was light, and what was its refraction, and fit the eyes of all creatures after the most curious manner to make use of it? These and suchlike considerations always have and ever will prevail with mankind to believe that there is a Being who made all things and has all things in his power, and who is therefore to be feared.[360]

> [T]he motions which the planets now have could not spring from any natural cause alone, but were impressed by an intelligent Agent. … To make this system, therefore, with all its motions, required a cause which understood and compared together the quantities of matter in the several bodies … and to compare and adjust all these things together, in so great a variety of bodies, argues that cause to be, not blind and fortuitous, but very well skilled in mechanics and geometry.[361]

> God made and governs the world invisibly and has commanded us to love and worship him…. And by the same power by which he gave life at first to every species of animal he is able to revive the dead, and has revived Jesus Christ our Redeemer….[362]

Newton's magnum opus is entitled *Mathematical Principles of Natural Philosophy* (or the *Principia* for short). Its contents exhibit such extraordinary brilliance that the famous French mathematician and astronomer Laplace referred to it as "pre-eminent above any other production of human genius." Likewise, the great Austrian physicist Ludwig Boltzmann called it "the first and greatest work ever written on theoretical physics."[363] Of this work, Newton declared:

> When I wrote my treatise about our system I had an eye upon such principles as might work with considering men for the belief of a Deity, and nothing can rejoice me more than to find it useful for that purpose.[364]

Another individual who is undoubtedly one of the most exceptional scientists ever to have lived is James Clerk Maxwell. Although not as well-known to the general public as Newton and Einstein, Maxwell left a legacy of accomplishments considered to be on the same plane of scientific greatness. He made many important contributions to diverse realms of physics, but most importantly, he discovered the fundamental laws that govern electricity, light, and magnetism.[365] [366] [367] What is the significance of this in practical terms? The formulas he derived "provided the tools to create everything from radar to radios and televisions to mobile phones."[368] Max Planck, founder of the discipline of quantum physics, asserted Maxwell "achieved greatness unequalled"[369] and referred to his laws of electromagnetism as one of the "greatest of all intellectual achievements."[370] The eminent physicist Richard Feynman affirmed:

> From a long view of the history of mankind … there can be little doubt that the most significant event of the nineteenth century will be judged as Maxwell's discovery of the laws of electrodynamics.[371]

This degree of acclaim from scientists of such caliber reveals Maxwell was no ordinary intellectual. We are not speaking of someone who was just book-smart but of an inspired visionary gifted with extraordinary intellect. Maxwell was also the antithesis of an eccentric genius lacking in social grace. According to his biographer, Maxwell was "generous, considerate, brave, genial, entertaining, and entirely without vanity or pretense."[372] What do we know of the spiritual views of this remarkable person? Listen to his own words:

> I have looked into most philosophical systems, and I have found that none will work without a God.[373]

> The belief in design is a necessary consequence of the Laws of Thought acting on the phenomena of perception.[374]

> Our flesh is God's making, who made us part of His world; but then He has given us the power of coming nearer to Himself, and so we ought to use the world and our bodies as means towards the knowledge of Him, and stretch always as far as our state will permit towards Him.[375]

> So we ought always to hope in Christ….[376]

> I believe, with the Westminster Divines and their predecessors *ad*

Infinitum that "Man's chief end is to glorify God and to enjoy him for ever."[377]

Take note that Maxwell was leery of linking scientific theories to the Bible, not for fear that the Bible might be wrong but because the theories of science were constantly changing, and he didn't want these fleeting ideas getting "fastened" to the Bible:

> The rate of change of scientific hypothesis is naturally much more rapid than that of Biblical interpretations so that if an interpretation is founded on such a hypothesis, it may help to keep the hypothesis above ground long after it ought to be buried and forgotten.[378]

Another outstanding scientist of Christian faith was George Washington Carver. Although barred from attending the elementary school where he lived because it only admitted white children, he rose above this and many other obstacles to become the man whom Henry Ford acknowledged as "the world's greatest living scientist."[379] Carver is generally known for developing hundreds of products derived from peanuts, but he is also responsible for many other important contributions to agriculture.[380] He pioneered the science of chemurgy, which is the utilization of organic materials for industrial purposes.[381] He was responsible for massive increases in farmland productivity and hybridized "whole families of fruits and plants" to make them "resistant to fungus attack."[382] With regard to his area of expertise, Carver had this to say:

> I love to think of nature as unlimited broadcasting stations, through which God speaks to us every day, every hour and every moment of our lives, if we will only tune in and remain so.[383]

> I am more and more convinced, as I search for truth that no ardent student of nature, can "Behold the lilies of the field"; or "Look unto the hills," or study even the microscopic wonders of a stagnant pool of water, and honestly declare himself to be an Infidel.[384]

The history of science is replete with momentous advancements made by scientists who were strongly convinced of the existence of God and the accuracy of the Bible. This list includes but is in no way restricted to such luminaries as William Thomson (also known as Lord Kelvin), who entered college at the age of ten and went on to publish more than 600 scientific papers. Among his many noteworthy accomplishments,[385] Thomson was a pioneer in the science of thermodynamics, which has proven to be among the most concrete and reliable of all scientific disciplines. Of it, Einstein said, "It is the only physical theory of universal content concerning which I am convinced that, within the framework of its basic concepts, it will never be overthrown."[386] The Second Law of Thermodynamics, which Thomson played a major role in developing,[387] was recognized by astrophysicist Arthur Eddington as holding "the supreme position among the laws of Nature."[388] In a letter to his wife, the vaunted physicist Helmholtz wrote that Thomson:

> far exceeds all the great men of science with whom I have made personal acquaintance, in intelligence, and lucidity, and mobility of thought, so that I felt quite wooden beside him sometimes.[389]

And what of William Thomson's views of spiritual matters? In 1903, before an audience at University College London, he declared:

> If you think strongly enough you will be forced by science to the belief in God, which is the foundation of all religion.[390] [391]

Among his many notable achievements, Louis Pasteur created the sciences of microbiology, stereochemistry, virology, immunology, and bacteriology.[392] [393] This legendary scientist also declared that science "brings man nearer to God."[394] He is mentioned only briefly here because we shall explore important consequences of his discoveries in an upcoming chapter.

Robert Boyle forged the discipline of modern chemistry and formulated what is known as Boyle's Law, which governs certain properties of gases. He was a prolific experimenter, and as the *World Book Encyclopedia* explains, "He helped establish the experimental method in chemistry and physics."[395] [396] [397] Of this method, Boyle wrote:

> And indeed, the experimental philosophy giving us a more clear discovery than strangers to it have of the divine excellencies displayed in the fabric and conduct of the universe and of the creatures it consists of, very much indisposes the mind to ascribe such admirable effects to so incompetent and pitiful a cause as blind chance, or the tumultuous jostlings of atomical portions of senseless matter; and leads it directly to the acknowledgment and adoration of a most intelligent, powerful and benign Author of things, to whom such excellent productions may with the greatest congruity be ascribed.[398]

Similarly, Boyle affirmed:

> And the more wonderful things [a person] discovers in the works
> of nature, the more auxiliary proofs he meets with to establish and
> enforce the argument, drawn from the Universe and its parts, to
> evince *that there is a God.*...[399]

> That this vast, beautiful, orderly and (in a word) many ways admi-
> rable system of things that we call the World, was framed by an
> Author supremely powerful, wise and good, can scarcely be denied
> by an intelligent and unprejudiced considerer.[400]

The famed astronomer and mathematician Johannes Kepler founded
the science of optics and "discovered the basic laws governing the motion
of the planets around the sun."[401] [402] The *Encyclopedia Americana* testifies
that this achievement "placed astronomy on modern foundations."[403]
Immanuel Kant singled out Kepler as "the most acute thinker ever
born."[404] As a book published by Oxford University Press explains, the
achievements of Galileo and Newton "depended critically on Kepler's
genius."[405] Most scientifically literate people know of Kepler, but how
many are aware of the fact that his writings abound with references to
God?[406] [407] As Kepler said, "I am earnest with my religion, I don't play with
it."[408] In the dedication to his first major scientific work, Kepler wrote:

> Here is treated the Book of Nature which is so highly praised by
> the Holy Scriptures. [The Apostle] Paul presents it to the heathens
> so that they may see God in it just as the sun can be observed in
> water or a mirror.[409]

In contrast to the dogma that belief in God is merely a dogma, Kepler
asserted:

> The greater the freedom of thought the more will faith be
> awakened in the sincerity of those who are devoted to scientific
> research.[410]

And what about scientists of today? Do any of them consider the
Bible to be authoritative? Absolutely. In fact, one such list of more than
100 individuals with doctoral degrees in various disciplines is readily available on the Internet.[411] These are not people who compartmentalize their
faith from their academics or twist the Scriptures to form strained interpretations to match their scientific viewpoints. Rather, they are scholars
who agree with the plain meaning of the Bible. Scientists with this view-

point are by no means a majority, but truth is not determined by a vote
—history is replete with examples wherein the majority was clearly wrong.
As Galileo wrote:

> [I]n the sciences the authority of thousands of opinions is not
> worth as much as one tiny spark of reason in an individual
> man.[412]

Dr. John Baumgardner has a Ph.D. in geophysics and space physics
from UCLA. In 1997, *U.S. News & World Report* described him as "the
world's pre-eminent expert in the design of computer models for geophysical convection, the process by which the Earth creates volcanoes, earth-
quakes, and the movement of the continental plates."[413] He has stated:

> Science has flowed from a Christian understanding of reality, a
> Christian understanding of God, and a Christian understanding
> of the natural world. In general I believe that science is legitimate,
> that it does reveal the glory of God, that it does confirm what the
> Scriptures say is valid and true.[414]

Dr. André Eggen has a Ph.D. in animal genetics from the Federal
Institute of Technology in Switzerland. He has stated:

> I find it marvelous to discover how God used the genetic codes to
> inscript life—so brilliant, so amazing.[415]

Dr. Andrew G. Bosanquet holds a Ph.D. in biochemistry from the
University of Canterbury (New Zealand) and is the director of Bath Cancer Research. The medical community has learned that cancer-fighting
drugs are not equally effective for all people, and patients are sometimes
placed on medications that do not work for them and lead to harmful side
effects. Thus, Dr. Bosanquet pioneered a procedure whereby cancer cells
are removed from each individual patient and combined in a test tube with
various cancer-fighting medicines to determine if they have potential to
work for that patient. Drugs that test well in this procedure are "8 times
more likely to be of benefit than test-resistant drugs." This approach in
fighting cancer is practical, humane, inexpensive, and used by doctors all
over the world.[416] [417] Dr. Bosanquet thinks the Bible is the word of God
and has written:

> I was increasingly convinced that God's guidance and help were
> necessary and available in every area of life; faith was not taking
> part in church services so much as following God's way through

each day. And this must include scientific research.[418]

So what is the point? Simply this—by the evidence shown in this chapter and in the judgment of some exceptionally gifted scientific minds, science and the Bible are not in conflict with one another but surprisingly harmonious. Given the length of the Bible and the fact that the most modern portions of it were written about two millennia ago, one would expect a myriad of scientific inaccuracies. What we find instead is the exact opposite. Is this the result of divine intervention? It would be quite a stretch to conclude this came about purely by chance.

Obviously, this doesn't mean a Christian or Christian institution is incapable of being completely wrong on a scientific matter. The clash between Galileo and the Catholic Church is a prime example of just such a case. However, one important fact frequently missing from descriptions of this famous episode in history should be pointed out: Galileo was also a Christian.[419] Moreover, he wrote that:

> the Bible can never speak untruth—whenever its true meaning is understood.[420]

CHAPTER 3

ARCHAEOLOGY

Archaeology is the scientific study of artifacts made by or associated with human beings.[421] Although this basic explanation sounds rather dry, this discipline and the light it sheds upon the Bible is fascinating. Similar to our study of ancient history in Chapter 1, we will cover some background information to acquire basic foundational knowledge in the subject. Instead of then analyzing several matters in depth, however, we will survey an expansive list of people, places, and events mentioned in the Bible and corroborated by archaeological evidence. Finally, opposing evidence will be discussed and the broad picture summarized.

DATING ARTIFACTS

Determining the age of artifacts is at the core of archaeological research. It is not enough to discover a relic or lost city. For such finds to be of any real historical value, they must be accurately dated. In this golden age of technology, one might think we can determine the age of any object with a laboratory test, but this is far from the case.[422] [423] While the invention of carbon-14 dating (also called radiocarbon or C-14 dating) is considered a great advance for archaeology,[424] [425] this method only works on objects that were once living.[426] Other laboratory tests can be used to date some inorganic materials, but such tests don't tell us when these materials came into use by humans and, for instance, were chiseled into arrowheads or bricks.

Even in situations in which carbon-14 dating is applicable, the accuracy limitations can be significant. For example, in 2003, a plant found in a plaster wall of a tunnel in Jerusalem was dated in this manner, and the result was that the calendar age of the plant was determined to be, "with 95% probability, between 800 and 510 BC."[427] Probabilities and ranges are a reality of scientific dating methods—uncertainties abound,[428] [429] but news accounts often spare the technical details, leaving readers with the perception that the results are more solid than they really are. For example, the C-14 test just cited was published in the scientific journal *Nature*, yet the news division of this very same company reported the date as "700-800 BC."[430]

Additionally, even when a laboratory test (or tests) can confidently date an object to within a range of 100 years, massive debates can rage over a disagreement of 75 years. This is not just petty bickering because a variance of 75 years can sometimes have massive implications for the conclusions we reach.[431]

Some artifacts such as coins can be dated fairly accurately by virtue of the fact that they were minted by rulers whose time in office can be established through historical texts. Since coins were plentiful in many ancient empires, the archaeological record is loaded with them, and researchers capitalize on this by using coins as a reference point to date other artifacts.[432] [433] For example, very recently, a structure mentioned in the New Testament (the Pool of Siloam) was uncovered. Excavators used a metal detector to discover some coins imbedded in the plaster walls of this structure. The coins were those of a king who ruled from 107–76 B.C., which in all probability tells us the pool was built around this time or shortly thereafter.[434] This kind of explicit evidence, however, is the exception not the rule.

More frequently, objects are dated through a process of inference that involves a greater degree of uncertainty. Scattered throughout modern-day Israel and Jordan are thousands of earth mounds where ancient cities, towns, or communities once stood. These are referred to as tels or tells.

http://en.wikipedia.org/wiki/File:Tel_hatzor.JPG

Tel Hatzor (also known Hazor)

Tels contain layers of civilization built on top of each another.[435] Amazingly, these mounds can be over a hundred feet high and can contain the remains of up to 20 cities.[436] This strange phenomenon is a result of the following factors:

1) Most ancient homes were made of non-durable materials that quickly disintegrated and were easily destroyed by natural disasters.

2) It was not uncommon for a city or town to be wiped out by an invading army.[437]

3) Building sites were chosen based upon water supply, defensibility, and convenience for trade and travel. Thus, locations with these qualities were continually reused.

4) Stone ruins were a handy source of construction materials, and it was easier to rebuild pre-existing outer defense walls than to start from scratch.[438] This made the sites of abandoned old settlements preferable locations for new ones.

When archaeologists excavate a tel, they attempt to date each successive layer of occupation. So when they find coins in a certain layer, they can use these as a reference point to date that layer.[439] [440] Pottery is also often used in this manner, as archaeologists have determined that certain styles and manufacturing methods were popular in different eras.[441] [442] However, such methods leave plenty of room for uncertainty. Pottery styles fell into and out of use at varying locations during different periods.[443] [444] Also, layers were disrupted and intermingled when the ancient inhabitants of these locations dug the ground for various purposes.[445] For instance, coins were often buried by their owners for safekeeping, and outer defense walls were frequently dug out for reuse.[446] [447]

It should be evident by now that the dating of archaeological remains is not an exact science, and we should be cautious about being too definitive in our conclusions.

FORGERIES

In addition to the vagaries of dating artifacts, archaeologists also face the problem of evaluating their authenticity. It has been said that a "forgery hysteria" is "consuming archaeological circles in Israel."[448] The Israel Antiquities Authority (IAA), a government agency tasked with enforcing laws and regulating matters pertaining to archaeological discoveries, has recently declared nearly 200 artifacts to be forgeries. Among these are some very notable and publicized items, including a limestone burial box that bears the inscription, "James, son of Joseph, brother of Jesus."[449] These names correspond to Jesus' family as described in the Bible.[450]

Despite the reports of some news outlets,[451] this is far from an open and shut case. Some prominent scholars have analyzed several of these objects and concluded they are genuine.[452] [453] [454] Others are undecided at this point.[455] [456] And some who have declared certain objects to be forgeries have later qualified this by saying they are 80–90% sure.[457] On top of this, the preliminary report published by the IAA regarding the burial box of James is rife with demonstrably false assertions.[458] Moreover, one of the scholars appointed by the IAA to analyze this box first concluded it was authentic, then decided it was a forgery, and later announced it was authentic.[459]

As we have seen, laboratory tests are not always as conclusive as they are made out to be. This applies not only to dating objects but also to evaluating their authenticity.[460] [461] This fact is vividly reinforced in this case, in which certain tests, or at least some scholars' interpretation of the test results, tell us the James burial box is genuine—while other scholars tell us the exact opposite based upon a different test.[462] [463] [464] For now, the case sits in an Israeli court, and we await more information.

How Can They Be So Sure?

Given what we have discussed thus far, I am often startled by the level of certainty expressed by archaeological scholars. For whatever reason, discourse among archaeologists tends to be pretty heated, and in such an environment, it may be that people state their case more surely than the evidence warrants. I would like to avoid this and am therefore qualifying the list below by stating that in my opinion, the level of certainty for each item is in the range of 90% or more. In a list of this length, it is likely time will reveal information that will cause me to change my opinion on some points. If this should happen, I will make note of it at rationalconclusions. com. Furthermore, it should be pointed out that every entry is within the scope of mainstream scholarly opinion.

Because this list does not delve into details, I have limited the citations to only two key sources so readers who would like more specifics can readily access them. The first is a book entitled *On the Reliability of the Old Testament* by Kenneth A. Kitchen. The author is an emeritus professor (one who is retired, but holds an honorary position) at the University of Liverpool's School of Archaeology, Classics and Egyptology. His specialties include the archaeology, history, and culture of ancient Egypt, the Near East, and the Old Testament.[465] [466] As with nearly every source I have cited, I don't agree with all the author writes, but this book is among the more impressive I have read for its mastery of facts and primary source data. The majority of entries below are supported by information found in this book.

The second source is the magazine *Biblical Archaeology Review*. This publication is notable for its comprehensible articles written by prominent scholars. While some archaeological journals have policies that result in the censorship of important information,[467] [468] *Biblical Archaeology Review* thrives on publishing opposing viewpoints from a variety of sources, and thus, readers are exposed to a diverse array of facts and opinions. This journal was instrumental in breaking the stranglehold that held up publication of various Dead Sea Scrolls for 30 years.[469] These two resources are abbreviated in the online citations as *OROT* and *BAR*. They are occasionally supplemented by information from the Encyclopædia Britannica. Also, I have included pictures of several coins I own that were minted by people mentioned in the Bible.

Inscriptions (writings that appear on ancient artifacts) are considered archaeological evidence and are therefore the basis of many items below.[470] In contrast, any objects that have been accused by the Israel Antiquities Authority of being a forgery are not the basis of any items in this list. Note that some allegedly forged artifacts are inscribed with names that also appear on artifacts that are not in question.[471] [472] Also note that ancient cities were sometimes abandoned for centuries, and thus, several cities are listed below more than once to show they were inhabited during separate important periods.[473]

In some academic circles, what follows is considered to be an outrage. Certain scholars want the term "biblical archaeology" stricken from academic literature. They say archaeology "has no business trying to prove" the Bible.[474] I say facts are facts, and it is censorship to restrict their use. Regardless, given the uncertainties we must contend with in archaeology, I think "prove" is too strong a word to use and prefer to say the following evidence corroborates the Bible.

Note that all entries must fit chronologically and geographically. In other words, the archaeological evidence must attest to more than just basic places, people, and events. It must also be compatible with the location and approximate timing of such places, people, and events as described in the Bible. While the following list is extensive, it is far from all-encompassing. For those who would prefer to skim this lengthy compilation, I have placed a star (*) in front of those items I consider most significant. Also, in four places, I have placed two stars to highlight items I consider extraordinarily significant.

Cities, Towns, Empires, and Landmarks in the Bible Corroborated by Archaeological Evidence

*Cities of Ur and Haran (2,000–1,800 B.C.)[475]

Genesis 11:31: "And Terah took Abram his son, and Lot the son of Haran his son's son, and Sarai his daughter in law, his son Abram's wife; and they went forth with them from Ur of the Chaldees, to go into the land of Canaan; and they came unto Haran, and dwelt there."

City of Shechem (2,000–1,550 B.C.)[476]

Genesis 33:18: "And Jacob came to Shalem, a city of Shechem, which *is* in the land of Canaan, when he came from Padanaram; and pitched his tent before the city."

Town of Dothan (18th/17th centuries B.C.)[477]

Genesis 37:14: "And the man said, They are departed hence; for I heard them say, Let us go to Dothan. And Joseph went after his brethren, and found them in Dothan."

*City of Hebron (17th century B.C.)[478]

Genesis 37:14: "So he [Israel] sent him [Joseph] out of the vale of Hebron…."

Lands of Edom and Moab (14th/13th centuries B.C.)[479]

Exodus 15:5: "Then the dukes of Edom shall be amazed; the mighty men of Moab, trembling shall take hold upon them…."

City of Jerusalem (14th/13th centuries B.C.)[480][481]

Joshua 10:1: "Now it came to pass, when Adonizedek king of Jerusalem…."

Town of Azekah (13th century B.C.)[482]

Joshua 10:10: "And the LORD discomfited them before Israel, and slew them with a great slaughter at Gibeon, and chased them along the way that goeth up to Bethhoron, and smote them to Azekah, and unto Makkedah."

Town of Beth-Shemesh (13th century B.C.)[483]

Judges 1:33: "Neither did Naphtali drive out the inhabitants of Beth-shemesh…."

Town of Libnah (13th century B.C.)[484]

Joshua 10:9: "Then Joshua passed from Makkedah, and all Israel with him, unto Libnah, and fought against Libnah…."

City of Lachish (13th century B.C.)[485]

Joshua 10:31: "And Joshua passed from Libnah, and all Israel with him, unto Lachish, and encamped against it, and fought against it…."

City of Gezer (13th century B.C.)[486]

Joshua 10:33: "Then Horam king of Gezer came up to help Lach-ish…."

Town of Eglon (13th century B.C.)[487]

Joshua 10:34: "And from Lachish Joshua passed unto Eglon…."

*City of Hazor (13th century B.C.)[488]

Joshua 11:10: "And Joshua at that time turned back, and took Hazor…."

City of Bethel (13th century B.C.)[489]

Joshua 12:16 mentions "the king of Bethel."

City of Aphek (13th century B.C.)[490]

Joshua 12:18 mentions "the king of Aphek."

City of Achshaph (13th century B.C.)[491]

Joshua 12:20 mentions "the king of Achshaph."

City of Hebron (13th century B.C.)[492]

Joshua 14:13: "And Joshua blessed him, and gave unto Caleb the son of Jephunneh Hebron for an inheritance."

City of Dibon (13th century B.C.)[493]

Numbers 32:34: "And the children of Gad built Dibon, and Ataroth, and Aroer…."

Land of Seir (13th century B.C.)[494]

Judges 5:4: "LORD, when thou wentest out of Seir…."

City of Taanach (13th/12th centuries B.C.)[495]

Joshua 12:21 mentions "the king of Taanach."

* City of Gath (1,200–600 B.C.)[496]

1 Samuel 17:4: "And there went out a champion out of the camp of the Philistines, named Goliath, of Gath, whose height *was* six cubits and a span."

City of Shechem (12th century B.C.)[497]

Judges 9:1: "And Abimelech the son of Jerubbaal went to Shechem unto his mother's brethren, and communed with them…."

City of Askelon (12th–11th centuries B.C.)[498]

2 Samuel 1:20: "Tell *it* not in Gath, publish *it* not in the streets of Askelon; lest the daughters of the Philistines rejoice, lest the daughters of the uncircumcised triumph."

City of Gaza (12th–11th centuries B.C.)[499]

1 Samuel 6:12: "And these *are* the golden emerods which the Philistines returned *for* a trespass offering unto the LORD; for Ashdod one, for Gaza one, for Askelon one, for Gath one, for Ekron one…."

City of Ekron (12th–11th centuries B.C.)[500]

1 Samuel 5:10: "Therefore they sent the ark of God to Ekron."

City of Beth-Shan (11th century B.C.)[501]

1 Samuel 31:10: "And they put his armour in the house of Ashtaroth: and they fastened his body to the wall of Bethshan."

*City of Ashdod (12th–10th centuries B.C.)[502]

1 Samuel 5:1: "And the Philistines took the ark of God, and brought it from Ebenezer unto Ashdod."

City of Shiloh (12th–11th centuries B.C.)[503]

Judges 21:12: "[A]nd they brought them unto the camp to Shiloh, which *is* in the land of Canaan."

*The Pool of Gibeon (11th century B.C.)[504]

2 Samuel 2:13: "And Joab the son of Zeruiah, and the servants of David, went out, and met together by the pool of Gibeon: and they sat down, the one on the one side of the pool, and the other on the other side of the pool."

Town of Beth-Shemesh (11th century B.C.)[505]

1 Samuel 6:9: "And see, if it goeth up by the way of his own coast to Bethshemesh…."

City of Dan (11th/10th centuries B.C.)[506]

1 Samuel 4:20: "And all Israel from Dan even to Beersheba knew that Samuel *was* established *to be* a prophet of the LORD."

City of Tadmor (11th/10th centuries B.C.)[507]

2 Chronicles 8:4: "And he [Solomon] built Tadmor in the wilderness, and all the store cities, which he built in Hamath."

City of Dor (11th–4th centuries B.C.)[508]

1 Kings 4:11: "The son of Abinadab, in all the region of Dor; which had Taphath the daughter of Solomon to wife…."

City of Gezer (10ᵗʰ century B.C.)[509]

1 Kings 9:15: "And this *is* the reason of the levy which king Solomon raised; for to build the house of the LORD, and his own house, and Millo, and the wall of Jerusalem, and Hazor, and Megiddo, and Gezer."

Town of Elon Beth-Hanan (10ᵗʰ century B.C.)[510]

1 Kings 4:7, 9: "And Solomon had twelve officers over all Israel…. The son of Dekar, in Makaz, and in Shaalbim, and Bethshemesh, and Elonbethhanan…."

*City of Jerusalem and 30 locations in the surrounding region (10ᵗʰ century B.C.)[511] [512] [513]

2 Samuel 5:5: "In Hebron he [David] reigned over Judah seven years and six months: and in Jerusalem he reigned thirty and three years over all Israel and Judah."

1 Kings 9:19: "And all the cities of store that Solomon had, and cities for his chariots, and cities for his horsemen, and that which Solomon desired to build in Jerusalem, and in Lebanon, and in all the land of his dominion."

Cities of Hazor and Megiddo (10ᵗʰ century B.C.)[514]

1 Kings 9:15: "And this *is* the reason of the levy which king Solomon raised; for to build the house of the LORD, and his own house, and Millo, and the wall of Jerusalem, and Hazor, and Megiddo, and Gezer."

Land of Cabul and 36 places located therein (10ᵗʰ–8ᵗʰ centuries B.C.)[515]

1 Kings 9:13: "And he said, What cities *are* these which thou hast given me, my brother? And he called them the land off Cabul unto this day."

Region of Samaria (9ᵗʰ–6ᵗʰ centuries B.C.)[516]

Jeremiah 23:13: "And I have seen folly in the prophets of Samaria; they prophesied in Baal, and caused my people Israel to err."

*Region of Ophir and the fact that it was a source of gold (8ᵗʰ century B.C.)[517]

Isaiah 13:12: "I will make a man more precious than fine gold; even a man than the golden wedge of Ophir."

Note: The passage above is one of ten in the Bible that associate the region of Ophir with gold. Other examples include:

- 1 Kings 22:48: "Jehoshaphat made ships of Tharshish to go to Ophir for gold: but they went not; for the ships were broken at Eziongeber." [Jehoshaphat reigned in 9th century B.C.]
- 1 Kings 9:7, 28: "And Hiram sent in the navy his servants, ship-men that had knowledge of the sea, with the servants of Solomon. And they came to Ophir, and fetched from thence gold, four hundred and twenty talents, and brought *it* to king Solomon." [Solomon reigned in the 10th century B.C.]

City of Ekron (8th–7th centuries B.C.)[518]

Zephaniah 2:4: "For Gaza shall be forsaken, and Ashkelon a desolation: they shall drive out Ashdod at the noon day, and Ekron shall be rooted up."

City of Zoan (8th century B.C.)[519] [520]

Isaiah 19:11: "Surely the princes of Zoan *are* fools, the counsel of the wise counsellors of Pharaoh is become brutish...."

City of Ashdod (8th century B.C.)[521]

Isaiah 20:1: "In the year that Tartan came unto Ashdod, (when Sargon the king of Assyria sent him,) and fought against Ashdod, and took it...."

Regions of Gilead and Galilee (8th century B.C.)[522]

2 Kings 15:29: "In the days of Pekah king of Israel came Tiglathpileser king of Assyria, and took Ijon, and Abelbethmaachah, and Janoah, and Kedesh, and Hazor, and Gilead, and Galilee, all the land of Naphtali, and carried them captive to Assyria."

City of Tirzah (12th–6th centuries B.C.)[523]

1 Kings 16:17: "And Omri went up from Gibbethon, and all Israel with him, and they besieged Tirzah."

Region of Cush (8th/7th centuries B.C.)[524] [525]

Isaiah 11:11: "And it shall come to pass in *that* day, that the Lord shall set his hand again the second time to recover the remnant of his people, which shall be left, from Assyria, and from Egypt, and from Pathros, and from Cush, and from Elam, and from Shinar, and from Hamath, and from the islands of the sea."

Kingdom of Assyria (9th–7th centuries B.C.)[526] [527]

2 Kings 15:19: "*And* Pul the king of Assyria came against the land…."

Town of Timnah (8th–7th centuries B.C.)[528]

2 Chronicles 28:18: "The Philistines also had invaded the cities of the low country, and of the south of Judah, and had taken Bethshemesh, and Ajalon, and Gederoth, and Shocho with the villages thereof, and Timnah with the villages thereof…."

City of Lachish (9th–7th centuries B.C.)[529]

2 Chronicles 32:9: "After this did Sennacherib king of Assyria send his servants to Jerusalem, (but he *himself laid siege* against Lachish, and all his power with him)…."

*Babylonian Empire (6th century B.C.)[530]

2 Kings 24:1 "In his days Nebuchadnezzar king of Babylon came up, and Jehoiakim became his servant three years: then he turned and rebelled against him."

Town of Tahpanhes (7th/6th century B.C.)[531] [532]

Jeremiah 43:7: "So they came into the land of Egypt: for they obeyed not the voice of the LORD: thus came they *even* to Tahpanhes."

Persian city of Susa (also spelled Shushan) and the palaces located there (6th–4th centuries B.C.)[533] [534]

Esther 1:2: "*That* in those days, when the king Ahasuerus sat on the throne of his kingdom, which *was* in Shushan the palace…."

Nehemiah 1:1: "The words of Nehemiah the son of Hachaliah. And it came to pass in the month Chisleu, in the twentieth year, as I was in Shushan the palace….”

City of Bethel (6th–4th centuries B.C.)535

Ezra 2:28: “The men of Bethel and Ai, two hundred twenty and three.”

Town of Gibeon (6th–4th centuries B.C.)536

Nehemiah 3:7: “And next unto them repaired Melatiah the Gibeonite, and Jadon the Meronothite, the men of Gibeon, and of Mizpah, unto the throne of the governor on this side the river.”

Town of Mizpah (6th–4th centuries B.C.)537

Nehemiah 3:19: “And next to him repaired Ezer the son of Jeshua, the ruler of Mizpah….”

City of Jerusalem (1st century A.D.)538

Luke 2:42: “And when he [Jesus] was twelve years old, they went up to Jerusalem after the custom of the feast.”

*Town of Capernaum (1st century A.D.)539

Matthew 8:5: “And when Jesus was entered into Capernaum, there came unto him a centurion, beseeching him….”

City/Region of Jericho (1st century A.D.)540

Matthew 20:29: “And as they [Jesus and his disciples] departed from Jericho, a great multitude followed him.”

**The Pool of Siloam (1st century A.D.)541

John 9:11: “He answered and said, A man that is called Jesus made clay, and anointed mine eyes, and said unto me, Go to the pool of Siloam, and wash: and I went and washed, and I received sight.”

*Town of Magdala (1st century A.D.)542

Matthew 15:39: “And he [Jesus] sent away the multitude, and took ship, and came into the coasts of Magdala.”

City of Gadara (1ˢᵗ century A.D.)⁵⁴³ ⁵⁴⁴

Luke 8:26: "And they [Jesus and his disciples] arrived at the country of the Gadarenes, which is over against Galilee."

City of Caesarea Philippi (also known as Banias) (1ˢᵗ century A.D.)⁵⁴⁵

Matthew 16:13: "When Jesus came into the coasts of Caesarea Philippi, he asked his disciples, saying, Whom do men say that I the Son of man am?"

City of Caesarea Maritima (1ˢᵗ century A.D.)⁵⁴⁶

Acts 23:33: "Who, when they came to Caesarea, and delivered the epistle to the governor, presented Paul also before him."

City of Corinth (1ˢᵗ century A.D.)⁵⁴⁷

Acts 18:1: "After these things Paul departed from Athens, and came to Corinth…."

Region of Achaia (1ˢᵗ century A.D.)⁵⁴⁸

2 Corinthians 1:1: "Paul, an apostle of Jesus Christ by the will of God, and Timothy *our* brother, unto the church of God which is at Corinth, with all the saints which are in all Achaia…."

City of Antioch (in Syria) (1ˢᵗ century A.D.)⁵⁴⁹

Acts 13:1: "Now there were in the church that was at Antioch certain prophets and teachers…."

*City of Ephesus (1ˢᵗ century A.D.)⁵⁵⁰

Acts 20:16: "For Paul had determined to sail by Ephesus…."

PEOPLE IN THE BIBLE CORROBORATED BY ARCHAEOLOGICAL EVIDENCE

*The Hittites (20ᵗʰ century B.C.)⁵⁵¹

Genesis 49:30: "In the cave that *is* in the field of Machpelah, which *is* before Mamre, in the land of Canaan, which Abraham bought with the field of Ephron the Hittite for a possession of a burying place."

*Population called "Israel" located in the land of Canaan (13th century B.C.)[552]

Joshua 22:10–11: "And when they came unto the borders of Jordan, that *are* in the land of Canaan, the children of Reuben and the children of Gad and the half tribe of Manasseh built there an altar by Jordan, a great altar to see to. And the children of Israel heard say, Behold, the children of Reuben and the children of Gad and the half tribe of Manasseh have built an altar over against the land of Canaan, in the borders of Jordan, at the passage of the children of Israel."

*The Philistines (12th century B.C.)[553]

Judges 16:4–5: "And it came to pass afterward, that he [Samson] loved a woman in the valley of Sorek, whose name was Delilah. And the lords of the Philistines came up unto her…."

The Midianites (12th/11th centuries B.C.)[554]

Judges 6:6: "And Israel was greatly impoverished because of the Midianites…."

**David (of "David and Goliath"), king of Judah and Israel (11th/10th centuries B.C.)[555]

2 Samuel 2:4: "And the men of Judah came, and there they anointed David king over the house of Judah."

2 Samuel 5:3–5: "So all the elders of Israel came to the king to Hebron; and king David made a league with them in Hebron before the LORD: and they anointed David king over Israel. David *was* thirty years old when he began to reign, *and* he reigned forty years. In Hebron he reigned over Judah seven years and six months: and in Jerusalem he reigned thirty and three years over all Israel and Judah."

Shishak I (also spelled Shishaq, Shushaq, Shoshenq), king of Egypt (10th century B.C.)[556]

1 Kings 11:40: "And Jeroboam arose, and fled into Egypt, unto Shishak king of Egypt, and was in Egypt until the death of Solomon."

Omri, king of Israel (9th century B.C.)[557] [558]

1 Kings 16:16: "And the people *that were* encamped heard say, Zimri

hath conspired, and hath also slain the king: wherefore all Israel made Omri, the captain of the host, king over Israel that day in the camp.”

*Ahab, king of Israel (9th century B.C.)[559]

1 Kings 16:28: “So Omri slept with his fathers, and was buried in Samaria: and Ahab his son reigned in his stead.”

Jehoram II, king of Israel (9th century B.C.)[560]

2 Kings 3:1: “Now Jehoram the son of Ahab began to reign over Israel in Samaria the eighteenth year of Jehoshaphat king of Judah, and reigned twelve years.”

Ahaziah II, king of Judah (9th century B.C.)[561]

2 Kings 10:13: “Jehu met with the brethren of Ahaziah king of Judah….”

Jehu, king of Israel (9th century B.C.)[562]

2 Kings 10:36: “And the time that Jehu reigned over Israel in Samaria *was* twenty and eight years.”

Mesha, king of Moab (9th century B.C.)[563]

2 Kings 3:4: “And Mesha king of Moab was a sheepmaster, and rendered unto the king of Israel an hundred thousand lambs, and an hundred thousand rams, with the wool.”

Hazael, king of Syria (9th century B.C.)[564]

2 Kings 8:29: “And king Joram went back to be healed in Jezreel of the wounds which the Syrians had given him at Ramah, when he fought against Hazael king of Syria.”

Jehoash, king of Judah (8th century B.C.)[565]

2 Kings 12:18: “And Jehoash king of Judah took all the hallowed things that Jehoshaphat, and Jehoram, and Ahaziah, his fathers, kings of Judah, had dedicated, and his own hallowed things, and all the gold *that was* found in the treasures of the house of the LORD, and in the king's house, and sent *it* to Hazael king of Syria: and he went away

from Jerusalem."

Benhadad III, king of Syria (8th century B.C.)[566]

2 Kings 13:24: "So Hazael king of Syria died; and Benhadad his son reigned in his stead."

Jeroboam II, king of Israel (8th century B.C.)[567]

2 Kings 14:23: "In the fifteenth year of Amaziah the son of Joash king of Judah Jeroboam the son of Joash king of Israel began to reign in Samaria, *and reigned* forty and one years."

Uzziah, king of Judah (8th century B.C.)[568]

2 Kings 15:32: "In the second year of Pekah the son of Remaliah king of Israel began Jotham the son of Uzziah king of Judah to reign."

*Tiglath-pileser III, king of Assyria (8th century B.C.)[569]

2 Kings 15:29: "In the days of Pekah king of Israel came Tiglathpileser king of Assyria, and took Ijon, and Abelbethmaachah, and Janoah, and Kedesh, and Hazor, and Gilead, and Galilee, all the land of Naphtali, and carried them captive to Assyria."

Pekah, king of Israel (8th century B.C.)[570]

2 Kings 15:27: "In the two and fiftieth year of Azariah king of Judah, Pekah the son of Remaliah began to reign over Israel in Samaria, *and reigned* twenty years."

Jotham (also spelled Jehotham), King of Judah (8th century B.C.)[571]

2 Kings 15:32: "In the second year of Pekah the son of Remaliah king of Israel began Jotham the son of Uzziah king of Judah to reign."

Menahem, king of Israel (8th century B.C.)[572]

2 Kings 15:20: "And Menahem exacted the money of Israel, *even* of all the mighty men of wealth, of each man fifty shekels of silver, to give to the king of Assyria."

Hoshea, king of Israel (8th century B.C.)[573]

2 Kings 15:30: "And Hoshea the son of Elah made a conspiracy against Pekah the son of Remaliah, and smote him, and slew him, and reigned in his stead, in the twentieth year of Jotham the son of Uzziah."

Ahaz, king of Judah (8th century B.C.)[574]

2 Kings 16:1: "In the seventeenth year of Pekah the son of Remaliah Ahaz the son of Jotham king of Judah began to reign."

Shalmaneser V, king of Assyria (8th century B.C.)[575]

2 Kings 18:9: "And it came to pass in the fourth year of king Hezekiah, which *was* the seventh year of Hoshea son of Elah king of Israel, *that* Shalmaneser king of Assyria came up against Samaria, and besieged it."

Sargon II, king of Assyria (8th century B.C.)[576]

Isaiah 20:1: "In the year that Tartan came unto Ashdod, (when Sargon the king of Assyria sent him,) and fought against Ashdod, and took it…."

Merodach-Baladan II (also spelled Berodach-Baladan), king of Babylon (8th century B.C.)[577]

2 Kings 20:12: "At that time Berodachbaladan, the son of Baladan, king of Babylon, sent letters and a present unto Hezekiah: for he had heard that Hezekiah had been sick."

Sennacherib, king of Assyria (8th/7th centuries B.C.)[578]

2 Kings 19:36: "So Sennacherib king of Assyria departed, and went and returned, and dwelt at Nineveh."

*Hezekiah, king of Judah (8th/7th centuries B.C.)[579]

2 Kings 16:20: "And Ahaz slept with his fathers, and was buried with his fathers in the city of David: and Hezekiah his son reigned in his stead."

*Tirhakah (also spelled Taharqa or Tirhaqah), king of Ethiopia (7th

century B.C.)[580]

Isaiah 37:9: "And he heard say concerning Tirhakah king of Ethiopia, He is come forth to make war with thee."

Esarhaddon, king of Assyria (7th century B.C.)[581]

2 Kings 19:36–37: "So Sennacherib king of Assyria departed, and went and returned, and dwelt at Nineveh. And it came to pass, … that Adrammelech and Sharezer his sons smote him with the sword: and they escaped into the land of Armenia. And Esarhaddon his son reigned in his stead."

Manasseh, king of Judah (7th century B.C.)[582]

2 Kings 21:1: "Manasseh *was* twelve years old when he began to reign, and reigned fifty and five years in Jerusalem."

Meshullam the scribe, his son Azaliah, and his grandson Shaphan (7th century B.C.)[583]

2 Kings 22:3: "And it came to pass in the eighteenth year of king Josiah, *that* the king sent Shaphan the son of Azaliah, the son of Meshullam, the scribe, to the house of the LORD…."

*Gemariah, son of Shaphan (7th century B.C.)[584] [585]

Jeremiah 36:10: "Then read Baruch in the book the words of Jeremiah in the house of the LORD, in the chamber of Gemariah the son of Shaphan the scribe, in the higher court, at the entry of the new gate of the LORD'S house, in the ears of all the people."

Ahiqam (also spelled Ahikam), son of Shaphan (7th century B.C.)[586]

2 Kings 22:12: "And the king commanded … Ahikam the son of Shaphan, and Achbor the son of Michaiah, and Shaphan the scribe…."

Hilkiah, high priest (7th century B.C)[587]

2 Kings 22:12: "And the king commanded Hilkiah the priest…."

Azariah, son of Hilkiah (7th century B.C.)[588]

1 Chronicles 6:13: "And Shallum begat Hilkiah, and Hilkiah begat Azariah…"

Jehucal, the son of Shelemiah (7th/6th centuries B.C.)[589]

Jeremiah 37:3: "And Zedekiah the king sent Jehucal the son of Shelemiah … to the prophet Jeremiah, saying, Pray now unto the LORD our God for us."

*Necho II, king of Egypt (7th/6th centuries B.C.)[590]

2 Chronicles 35:20: "After all this, when Josiah had prepared the temple, Necho king of Egypt came up to fight against Carchemish by Euphrates: and Josiah went out against him."

*Nebuchadrezzar II, king of Babylon (7th/6th centuries B.C.)[591]

2 Kings 24:11: "And Nebuchadnezzar king of Babylon came against the city, and his servants did besiege it."

Pedaiah, son of Jeconiah (7th/6th centuries B.C.)[592]

1 Chronicles 3:16–18: "And the sons of Jehoiakim: Jeconiah his son, Zedekiah his son. And the sons of Jeconiah; Assir, Salathiel his son, Malchiram also, and Pedaiah, and Shenazar, Jecamiah, Hoshama, and Nedabiah."

Jehoiachin (also spelled Jehoiakim), king of Judah (6th century B.C.)[593]

2 Kings 24:1: "In his days Nebuchadnezzar king of Babylon came up, and Jehoiakim became his servant three years: then he turned and rebelled against him."

Pharaoh Hophra, king of Egypt (6th century B.C.)[594 595]

Jeremiah 44:30: "Thus saith the LORD; Behold, I will give Pharaohhophra king of Egypt into the hand of his enemies…."

Evil-Merodach (also spelled Awel-Marduk), king of Babylon (6th century B.C.)[596]

2 Kings 25:27: "And it came to pass in the seven and thirtieth year of the captivity of Jehoiachin king of Judah, in the twelfth month, on the seven and twentieth *day* of the month, *that* Evilmerodach king of Babylon in the year that he began to reign did lift up the head of Jehoiachin king of Judah out of prison…."

*Belshazzar, effective ruler of Babylon (6th century B.C.)[597]

Daniel 5:29: "Then commanded Belshazzar, and they clothed Daniel with scarlet, and *put* a chain of gold about his neck, and made a proclamation concerning him, that he should be the third ruler in the kingdom."

*Cyrus II, king of Persia (6th century B.C.)[598]

Ezra 1:1: "Now in the first year of Cyrus king of Persia…."

*Darius the Great, king of Persia (6th/5th centuries B.C.)[599][600]

Daniel 6:9: "Wherefore king Darius signed the writing and the decree."

Baalis, king of Ammon (6th century B.C.)[601]

Jeremiah 40:14: "And said unto him, Dost thou certainly know that Baalis the king of the Ammonites hath sent Ishmael the son of Nethaniah to slay thee? But Gedaliah the son of Ahikam believed them not."

Sanballat I, governor of Samaria (5th century B.C.)[602]

Nehemiah 2:10: "When Sanballat the Horonite, and Tobiah the servant, the Ammonite, heard *of it*, it grieved them exceedingly that there was come a man to seek the welfare of the children of Israel."

Herod, king of Judea (1st century B.C.)[603]

Matthew 2:1: "Now when Jesus was born in Bethlehem of Judaea in the days of Herod the king…."

Augustus Caesar, emperor of the Roman Empire (1st century B.C.)[604]

Luke 2:1: "And it came to pass in those days, that there went out a decree from Caesar Augustus, that all the world should be taxed."

Coin minted by and bearing the image of Augustus Caesar

*Archelaus, son of Herod (1ˢᵗ century A.D.)[605]

Matthew 2:22: "But when he [Joseph] heard that Archelaus did reign in Judaea in the room of his father Herod, he was afraid to go thither: notwithstanding, being warned of God in a dream, he turned aside into the parts of Galilee...."

Coin minted by Herod Archelaus

Philip, son of Herod (1ˢᵗ century A.D.)[606]

Luke 3:1: "[H]is brother Philip tetrarch of Ituraea and of the region of Trachonitis...."

Tiberius Caesar, emperor of the Roman Empire (1ˢᵗ century A.D.)[607]

Luke 3:1: "Now in the fifteenth year of the reign of Tiberius Caesar...."

Coin minted by and bearing the image of Tiberius Caesar

*Pontius Pilate, Roman governor of Judea (1ˢᵗ century A.D.)[608]

Matthew 27:2: "And when they had bound him [Jesus], they led *him* away, and delivered him to Pontius Pilate the governor."

Herod Agrippa II, tetrarch of Batanaea and Trachonitis (1ˢᵗ century A.D.)[609]

Acts 26:1: "Then Agrippa said unto Paul, Thou art permitted to speak for thyself."

*Erastus, city official in Corinth (1ˢᵗ century A.D.)[610]

Romans 16:23: "Erastus the chamberlain of the city saluteth you…."

2 Timothy 4:20: "Erastus abode at Corinth…."

CUSTOMS AND PRACTICES IN THE BIBLE CORROBORATED BY ARCHAEOLOGICAL EVIDENCE

*Worship of a god named Asherah (also spelled Asherat) (13ᵗʰ century B.C.)[611]

Deuteronomy 16:21 (NIV): "Do not set up any wooden Asherah pole beside the altar you build to the LORD your God…."

*Worship of a god named Baal (13ᵗʰ century B.C.)[612][613]

Numbers 22:41: "And it came to pass on the morrow, that Balak took Balaam, and brought him up into the high places of Baal, that thence he might see the utmost *part* of the people."

Judges 2:13: "And they forsook the LORD, and served Baal…."

*Pigs were not eaten in areas inhabited by Israelites but were in surrounding areas. (12ᵗʰ century B.C.)[614]

Leviticus 11:4, 7: "Nevertheless these shall ye not eat … the swine…."

Pharaohs giving their daughters in marriage to foreigners (10ᵗʰ–8ᵗʰ centuries B.C.)[615][616]

1 Kings 9:16: "*For* Pharaoh king of Egypt had gone up, and taken Gezer, and burnt it with fire, and slain the Canaanites that dwelt in the city, and given it *for* a present unto his daughter, Solomon's wife."

Worship of a goddess named Ashtoreth (also spelled Astarte) (10th century B.C.)[617]

1 Kings 11:4: "For Solomon went after Ashtoreth the goddess of the Zidonians…."

Note: Ashtoreth is sometimes confused with Asherah (also spelled Asherat) mentioned above.

*Babylonian practice of deporting the most skilled and learned citizens of a defeated enemy to Babylon and placing them in service of the Babylonian king (8th–6th centuries B.C.)[618]

Daniel 1:1–4: "In the third year of the reign of Jehoiakim king of Judah came Nebuchadnezzar king of Babylon unto Jerusalem, and besieged it. And the Lord gave Jehoiakim king of Judah into his hand…. And the king spake unto Ashpenaz the master of his eunuchs, that he should bring *certain* of the children of Israel, and of the king's seed, and of the princes; Children in whom *was* no blemish, but well favoured, and skilful in all wisdom, and cunning in knowledge, and understanding science, and such as *had* ability in them to stand in the king's palace…."

Babylonian practice of giving such captive servants special allowances (8th–6th centuries B.C.)[619]

Daniel 1:5: "And the king [Nebuchadnezzar] appointed them a daily provision of the king's meat, and of the wine which he drank: so nourishing them three years, that at the end thereof they might stand before the king."

2 Kings 25:27–30: "Evilmerodach king of Babylon in the year that he began to reign did lift up the head of Jehoiachin king of Judah out of prison; And he spake kindly to him, and set his throne above the throne of the kings that were with him in Babylon; And changed his prison garments: and he did eat bread continually before him all the days of his life. And his allowance *was* a continual allowance given him of the king, a daily rate for every day, all the days of his life."

*After conquering Babylon, Cyrus II (king of Persia), returned Babylonian captives to their native lands, helped rebuild their ruined temples, and returned sacred objects that had been taken from these temples. (6th century B.C.)[620] [621]

Ezra 1:1–3, 7–8, 11: "Now in the first year of Cyrus king of Persia … he made a proclamation throughout all his kingdom, and *put it* also in writing, saying, Thus saith Cyrus king of Persia, The LORD God of heaven hath given me all the kingdoms of the earth; and he hath charged me to build him an house at Jerusalem, which *is* in Judah. Who *is there* among you of all his people? his God be with him, and let him go up to Jerusalem, which *is* in Judah, and build the house of the LORD God of Israel, (he *is* the God,) which *is* in Jerusalem. … Also Cyrus the king brought forth the vessels of the house of the LORD, which Nebuchadnezzar had brought forth out of Jerusalem, and had put them in the house of his gods; Even those did Cyrus king of Persia bring forth … and numbered them unto Sheshbazzar, the prince of Judah. … All *these* did Sheshbazzar bring up with *them of* the captivity that were brought up from Babylon unto Jerusalem."

Persian authorities issued letters of safe passage (5th century B.C.)[622]

Nehemiah 2:7–8: "Moreover I said unto the king, If it please the king, let letters be given me to the governors beyond the river, that they may convey me over till I come into Judah…. And the king granted me, according to the good hand of my God upon me."

**Rich people used tombs that were cut into bedrock. (1st century B.C.)[623]

Matthew 27:57–60: "When the even was come, there came a rich man of Arimathaea, named Joseph, who also himself was Jesus' disciple: He went to Pilate, and begged the body of Jesus. Then Pilate commanded the body to be delivered. And when Joseph had taken the body, he wrapped it in a clean linen cloth, And laid it in his own new tomb, which he had hewn out in the rock: and he rolled a great stone to the door of the sepulchre, and departed."

Worship of a goddess named Diana in the Greek city of Ephesus (1st century A.D.)[624]

Acts 19:35: "And when the townclerk had appeased the people, he [Alexander] said, *Ye* men of Ephesus, what man is there that knoweth not how that the city of the Ephesians is a worshipper of the great goddess Diana…."

Events in the Bible Corroborated by Archaeological Evidence

*Shishak I (king of Egypt) attacked various places in Judea. (10th century B.C.)[625]

1 Kings 14:25: "And it came to pass in the fifth year of king Rehoboam, *that* Shishak king of Egypt came up against Jerusalem: And he took away the treasures of the house of the LORD, and the treasures of the king's house; he even took away all: and he took away all the shields of gold which Solomon had made."

War between Mesha (king of Moab) and Israel (9th century B.C.)[626]

2 Kings 3:4–5, 24: "And Mesha king of Moab was a sheepmaster, and rendered unto the king of Israel an hundred thousand lambs, and an hundred thousand rams, with the wool. But it came to pass, when Ahab was dead, that the king of Moab rebelled against the king of Israel. … And when they came to the camp of Israel, the Israelites rose up and smote the Moabites, so that they fled before them: but they went forward smiting the Moabites, even in *their* country."

Israelites adopted the worship of the foreign deity Asherah (also spelled Asherat). (9th/8th century B.C.)[627]

1 Kings 16:33 (NIV): "Ahab [king of Israel] also made an Asherah pole and did more to provoke the LORD, the God of Israel, to anger than did all the kings of Israel before him."

*Jehoram II (king of Israel) and Ahaziah II (king of Judah) killed together (9th century B.C.)[628]

2 Kings 9:24, 27: "And Jehu drew a bow with his full strength, and smote Jehoram between his arms, and the arrow went out at his heart, and he sunk down in his chariot. … But when Ahaziah the king of Judah saw *this*, he fled by the way of the garden house. And Jehu followed after him, and said, Smite him also in the chariot. *And they did so* at the going up to Gur, which *is* by Ibleam. And he fled to Megiddo, and died there."

*Earthquake in Israel between 776–750 B.C.[629]

Amos 1:1: "The words of Amos, who was among the herdmen of Tekoa, which he saw concerning Israel in the days of Uzziah king of Judah, and in the days of Jeroboam the son of Joash king of Israel, two years before the earthquake."

Ahaz (king Of Judah) paid tribute money to Tiglath-Pileser III (king of Assyria). (about 734 B.C.)[630]

2 Kings 16:7–8: "So Ahaz sent messengers to Tiglathpileser king of Assyria, saying, I *am* thy servant and thy son: come up, and save me out of the hand of the king of Syria, and out of the hand of the king of Israel, which rise up against me. And Ahaz took the silver and gold that was found in the house of the LORD, and in the treasures of the king's house, and sent *it for* a present to the king of Assyria."

Tiglath-Pileser III seized regions of Gilead and Galilee and deported the population. (8th century B.C.)[631]

2 Kings 15:29: "In the days of Pekah king of Israel came Tiglathpileser king of Assyria, and took Ijon, and Abelbethmaachah, and Janoah, and Kedesh, and Hazor, and Gilead, and Galilee, all the land of Naphtali, and carried them captive to Assyria."

*Sargon II (king of Assyria) sent a high-ranking official to attack the city of Ashdod, and he defeated it. (8th century B.C.)[632]

Isaiah 20:1: "In the year that Tartan [a high-ranking official[633]] came unto Ashdod, (when Sargon the king of Assyria sent him,) and fought against Ashdod, and took it…."

Shalmaneser V (king of Assyria) attacked the region of Samaria. (8th century B.C.)[634]

2 Kings 17:3, 5: "Against him came up Shalmaneser king of Assyria…. Then the king of Assyria came up throughout all the land, and went up to Samaria, and besieged it three years."

Sennacherib (king of Assyria) attacked numerous towns and cities in Judah including the city of Lachish. (about 701 B.C.)[635]

2 Kings 18:13: "Now in the fourteenth year of king Hezekiah did Sennacherib king of Assyria come up against all the fenced cities of Judah,

and took them."

2 Chronicles 32:9: "After this did Sennacherib king of Assyria send his servants to Jerusalem, (but he *himself laid siege* against Lachish, and all his power with him)...."

Hezekiah (king of Judah) paid monetary tribute to Sennacherib. (about 701 B.C.)[636]

2 Kings 18:14–16: "And Hezekiah king of Judah sent to the king of Assyria to Lachish, saying, I have offended; return from me: that which thou puttest on me will I bear. And the king of Assyria appointed unto Hezekiah king of Judah three hundred talents of silver and thirty talents of gold. And Hezekiah gave *him* all the silver that was found in the house of the LORD, and in the treasures of the king's house. At that time did Hezekiah cut off *the gold from* the doors of the temple of the LORD, and *from* the pillars which Hezekiah king of Judah had overlaid, and gave it to the king of Assyria."

*Sennacherib's troops surrounded Jerusalem, but no battle took place. (about 701 B.C.)[637]

2 Kings 18:17; 19:35–36: "And the king of Assyria sent ... a great host against Jerusalem. And it came to pass that night, that the angel of the LORD went out, and smote in the camp of the Assyrians an hundred fourscore and five thousand: and when they arose early in the morning, behold, they *were* all dead corpses. So Sennacherib king of Assyria departed, and went and returned, and dwelt at Nineveh."

Sennacherib killed by his sons (7th century B.C.)[638]

Isaiah 37:37–38: "So Sennacherib king of Assyria departed, and went and returned, and dwelt at Nineveh. And it came to pass, as he was worshipping in the house of Nisroch his god, that Adrammelech and Sharezer his sons smote him with the sword; and they escaped into the land of Armenia: and Esarhaddon his son reigned in his stead."

*Hezekiah (king of Judah) constructed a conduit for the city's water supply. (about 701 B.C.)[639]

2 Chronicles 32:30: "This same Hezekiah also stopped the upper watercourse of Gihon, and brought it straight down to the west side of

the city of David."

2 Kings 20:20: "And the rest of the acts of Hezekiah, and all his might, and how he made a pool, and a conduit, and brought water into the city, *are* they not written in the book of the chronicles of the kings of Judah?"

*Nebuchadnezzar (king of Babylon) defeated Jehoiachin (king of Jerusalem), carried him off to Babylon, and appointed someone in his stead. (about 597 B.C.)[640]

2 Kings 24:11, 15, 17: "And Nebuchadnezzar king of Babylon came against the city, and his servants did besiege it. … And he carried away Jehoiachin to Babylon, and the king's mother, and the king's wives, and his officers, and the mighty of the land, *those* carried he into captivity from Jerusalem to Babylon. … And the king of Babylon made Mattaniah his father's brother king in his stead, and changed his name to Zedekiah."

**While captive in Babylon, the sons of Jehoiakim (king of Judah) received an allowance from Nebuchadnezzar (king of the Babylonian Empire). (about 592 B.C.)[641]

Daniel 1:1, 3, 5: "In the third year of the reign of Jehoiakim king of Judah came Nebuchadnezzar king of Babylon unto Jerusalem, and besieged it. … And the king spake unto Ashpenaz the master of his eunuchs, that he should bring *certain* of the children of Israel, and of the king's seed, and of the princes…. And the king appointed them a daily provision of the king's meat, and of the wine which he drank…."

The Limits of What We Know

While bearing in mind the extent of this list, consider how much evidence has been obliterated over time. As mentioned earlier, millennia of natural decay, disasters, and wars took a heavy toll on the physical structures of ancient civilizations. To cite just one example out of many, in 14/15 A.D., the Roman Emperor Tiberius ordered thousands of troops to burn, kill, and flatten everything in their path for fifty miles.[642] [643] Even objects durable enough to survive such destruction (like building stones) were often recycled into newer structures, thus removing them from their original context and lessening the knowledge that can be gained from them.[644] [645]

Beyond the ravages of war, ancient building projects have destroyed much archaeological evidence, particularly in Jerusalem, where massive reconstruction has taken place throughout the ages. In some cases, land was cleared all the way down to bedrock in order to establish a firm foundation for new structures.[646] [647] [648] Even in modern times, construction projects have mutilated some tremendously important archaeological sites.[649] [650] [651] [652] Furthermore, hundreds of sites are damaged every year by looters, resulting in the loss of "entire pages of human and cultural history."[653]

On top of all this, only a minuscule fraction of what has survived has been uncovered. There are about 30,000 locations in Israel where antiquities are known to exist,[654] but as we will see, the vast majority of evidence at these sites remains concealed.

More than 40 years ago, the Israeli government formed an association to conduct a survey of known archaeological sites. This entails mapping the terrain, taking photographs, and collecting artifacts (primarily those that lie exposed above the ground). Such surveys are the "least expensive and quickest" way of acquiring archaeological data. Yet, as of 2009, only about one quarter of the land area in Israel has been surveyed.[655] [656] Also, more sites are being found all the time. For instance, farmers and construction workers often happen upon such artifacts.[657] [658]

Moreover, the process of surveying usually doesn't reach into the ground where most archaeological evidence is located. Digging below the surface requires excavation, and generally, this isn't done with a bulldozer. Most digging is done carefully by hand in order to prevent evidence from being destroyed.[659] This obviously takes considerable time. Besides the fact that there are limited financial resources for such work, the political and religious strife that besets the Middle East has impeded progress at various times.[660] Thus, an unbelievable number of known antiquity sites have not yet been excavated. An inquiry to the Israel Antiquities Authority requesting the latest figures was not answered,[661] but as of 1976, five thousand major ruins had been identified in the countries of Israel and Jordan, and merely 2% of these had been subjected to any kind of excavation. Additionally, of this 2%, only 30 had been the site of a major excavation.[662]

Furthermore, even major excavations are far from comprehensive. Over a hundred major excavations have been conducted in Jerusalem,[663] yet only about 5% of the city has been excavated.[664] Some prime sites that archaeologists would love to dig are off limits because they are in the vicinity of holy sites or simply because modern housing is located there.[665] [666] This is not unique to Jerusalem.[667] Rarely is more than 10% of any location excavated.[668]

In addition to the many factors severely limiting the amount of evidence at hand, consider that only a portion of what is available has been studied and published. This is mostly due to a lack of resources. For example, a writing system called "cuneiform" was the most prevalent script in the ancient Middle East.[669] Somewhere around 500,000 tablets containing cuneiform inscriptions have been unearthed. Yet, of these, "only a fraction have been studied."[670]

© iStockphoto.com/fotofrog

2,300-year-old clay tablet with cuneiform script

A chronic and significant problem in archaeology is the failure of scholars to properly record, analyze, and publish the results of their excavations.[671] [672] [673] [674] As a result, much evidence removed from the ground has yet to see the light of day, and some of it never will.

For example, a major excavation was conducted in Jerusalem between 1961 and 1967 by one of the world's foremost archaeologists. A decade later, she passed on without having published a final report. This job was handed over to other scholars, who took up her work in 1982. However, the field notes from the 1960s excavation were sloppy and incomplete. As a result of this (and probably other factors I am unaware of), two of the six volumes comprising the final report were still unpublished as of early 2006. Moreover, some of the discovered artifacts were thrown out, which, combined with poor excavation records, made full recovery of the evidence impossible.[675] [676] [677]

If you are thinking that what I have just described is an anomaly, think again. Between 1961 and 1985, there were four large-scale excavations in Jerusalem. In every case, the scholars leading the excavations passed on before they had written their final reports, and as of 1998, none of these reports were finished.[678] To make matters worse, a scholar or institution that takes up the task of completing such research often has "exclusive access" to the evidence gathered from these excavations. This can leave (and, indeed, has left) the rest of academia and the general public in the dark for long expanses of time.[679] [680]

The Irrationality of Arguments from Silence

The principal source of fresh water for ancient Jerusalem was a spring named Gihon.[681] [682] We know from historical records and archaeological evidence that Jerusalem was inhabited throughout the Byzantine era (from 330 to 1450 A.D.).[683] [684] [685] [686] Yet, according to a scholar with the Israel Antiquities Authority, excavations near the Gihon Spring have failed to turn up any artifacts from this 1,200-year period. He says this is one of those things that just can't be explained.[687]

I say the explanation is obvious. We only have access to what is published, which is a portion of what has been excavated, which is a tiny percentage of what has been surveyed, which is a slice of what has survived, which is a remnant of what has not been destroyed by modern construction, looters, and botched excavations, which is an infinitesimal fraction of what has not been wiped out by millennia of decay, warfare, building projects, and natural disasters.

A common saying among archaeologists is, "Absence of evidence, is not evidence of absence." Given everything above, this makes good sense. Can you imagine the absurdity of drawing definitive conclusions based upon what archaeology has not revealed thus far? Yet, some scholars do just this. Simply put, such reasoning is illogical and unconvincing.

Conflicts between Archaeological Evidence and the Bible

Several of the items in the lists above pertain to an Assyrian king named Sennacherib. In addition to what the Bible says of him, we know about this ruler through inscriptions that describe his military and building accomplishments. These comprise what are referred to as Sennacherib's *Annals*.

▶As military campaigns were completed, Sennacherib drew up new editions of his *Annals* to include the most recent information.[688] [689] For our purposes, the variants in these editions are insignificant. The same is true of the various translations made by scholars. To be consistent, I have quoted from the same edition and translation throughout. In addition, to be comprehensive, the online citations contain an alternate edition and an alternate translation for the key passage shown below.◀

With regard to events that transpired between Sennacherib and the Jewish king Hezekiah, Sennacherib and the Bible agree with each other on several points, but there are also some clear divergences. Here is Sennacherib's account:

> As for Hezekiah, the Jew, who did not submit to my yoke, 46 of his strong, walled cities, as well as the small cities in their neighborhood which were without number.... I besieged and took.... Himself, like a caged bird I shut up in his royal city of Jerusalem. Earthworks I threw up against him,—one coming out of the city-gate, I turned back to his misery. ... As for Hezekiah, the terrifying splendor of my majesty overcame him, and ... in addition to the 30 talents of gold and 800 talents of silver, (there were) ... his daughters, his harem, his male and female musicians, (which) he had (them) bring after me to Nineveh my royal city. To pay tribute ... he dispatched his messengers.[690] [691] [692]

Sennacherib's *Annals*

Courtesy of the Oriental Institute of the University of Chicago

Here is the Biblical account:

> In the fourteenth year of King Hezekiah's reign, Sennacherib king of Assyria attacked all the fortified cities of Judah and captured them. So Hezekiah king of Judah sent this message to the king of Assyria at Lachish: "I have done wrong. Withdraw from me, and I will pay whatever you demand of me." The king of Assyria exacted from Hezekiah king of Judah three hundred talents of silver and thirty talents of gold. So Hezekiah gave him all the silver that was found in the temple of the LORD and in the treasuries of the royal palace. ...

The king of Assyria sent his supreme commander, his chief officer and his field commander with a large army, from Lachish to King Hezekiah at Jerusalem. ... Then the commander stood and called out in Hebrew: "Hear the word of the great king, the king of Assyria! This is what the king says: Do not let Hezekiah deceive you. He cannot deliver you from my hand. Do not let Hezekiah persuade you to trust in the LORD when he says, 'The LORD will surely deliver us; this city will not be given into the hand of the king of Assyria.' Do not listen to Hezekiah. This is what the king of Assyria says: Make peace with me and come out to me ... until I come and take you to a land like your own.... Choose life and not death!" ...

And Hezekiah prayed to the LORD: "O LORD, God of Israel, enthroned between the cherubim, you alone are God over all the kingdoms of the earth. You have made heaven and earth. Give ear, O LORD, and hear; open your eyes, O LORD, and see; listen to the words Sennacherib has sent to insult the living God. ... Now, O LORD our God, deliver us from his hand, so that all kingdoms on earth may know that you alone, O LORD, are God." ... That night the angel of the LORD went out and put to death a hundred and eighty-five thousand men in the Assyrian camp. When the people got up the next morning—there were all the dead bodies! So Sennacherib king of Assyria broke camp and withdrew. He returned to Nineveh and stayed there.[693]

To recap, here are the main agreements:

1) Sennacherib conquered numerous towns and cities in Judah.

2) Hezekiah paid monetary tribute to Sennacherib.

3) Sennacherib's troops surrounded Jerusalem, but no battle took place.

And here are the main disagreements:

1) The Bible claims Hezekiah paid tribute after Sennacherib captured his cities but before Sennacherib's troops surrounded Jerusalem. Sennacherib claims Hezekiah did not pay tribute until after his troops surrounded Jerusalem. Also, the amount of the tribute differs: Sennacherib says it was "30 talents of gold and 800 talents of silver," and the Bible says "three hundred talents of silver and thirty talents of gold."

2) The Bible claims Sennacherib withdrew his troops from Jerusalem after an angel of God slew vast numbers of them. Sennacherib is

unclear regarding when his troops left Jerusalem and claims Hezekiah dispatched messengers to pay tribute to him at his capital city of Nineveh.

Don't jump to the conclusion that Sennacherib's version of events is more credible because there is no miracle involved. Even if we disregard the Biblical account, some vexing questions remain regarding Sennacherib's report. As a result, scholars have expended much effort trying to figure out what happened here.[694] [695] [696] The first irregularity is that Sennacherib was not the kind of person to leave Hezekiah in power with his capital city unharmed. In the same inscription quoted from above, Sennacherib makes clear how he deals with his foes:

> The warriors of Hirimme, wicked enemies, I cut down with the sword. Not one escaped. Their corpses I hung on stakes, surrounding the city (with them).

> Over the whole of his wide land I swept like a hurricane…. I besieged, I captured, I destroyed, I devastated, I burned with fire.

> [C]ities … who had not speedily bowed in submission at my feet, I besieged, I conquered, I carried off their spoil….

> I drew near to Ekron and slew the governors and nobles who had committed sin (that is, rebelled), and hung their bodies on stakes around the city.

> All of their bodies I bored through…. I cut their throats…. I made (the contents of) their gullets and entrails run down upon the wide earth…. I cut out their privates…."

These quotes are a sampling from more than 15 instances in which Sennacherib claims to have dealt harshly with his adversaries.[697] With the exception of Hezekiah, all who did not submit immediately to Sennacherib lost their homeland or their lives. The relatively soft penalty for Hezekiah is even more senseless in light of the fact that he was near the top of Sennacherib's enemies list.[698]

Furthermore, the manner in which Sennacherib claims to have accepted tribute is also suspect. If his troops had Jerusalem surrounded, why didn't he demand tribute right there from Hezekiah himself? Sennacherib claims to have made other rulers bring tribute before him and kiss his feet. Why, in contrast, would he allow Hezekiah to send messengers to deliver it afterwards? This account is not only strange for Sennacherib but also contrary to the practices of all Assyrian monarchs.[699 700]

For these reasons and others, many scholars have expressed doubts about Sennacherib's account. A critical examination of it implies Sennacherib may have suffered a defeat, which is exactly what the Bible states. One thing is certain: if he ever met with failure, Sennacherib was not the kind of person to inscribe a record of it in stone for future generations. Nothing drives this point home like the opening proclamation in his *Annals*:

> Sennacherib, the great king, the mighty king, king of the universe … perfect hero; mighty man … the powerful one who consumes the insubmissive, who strikes the wicked with the thunderbolt….[701]

Suffice it to say that Sennacherib's inscription is not compelling evidence against the Biblical account. In fact, using the same standard as in the list above (90% certainty), I have yet to encounter archaeological evidence that shows any part of the Bible to be inaccurate. Nor am I alone in this assessment.[702]

One of the greatest archaeologists of the 20th century was a college president and rabbi by the name of Nelson Glueck. His extraordinary accomplishments include the discovery of approximately 1,500 archaeological sites in the lands of the Bible. He was featured on the cover of *Time* magazine for his archaeological endeavors, and when John F. Kennedy was sworn in as President of the United States, Glueck delivered the benediction.[703 704] Based upon decades of archaeological research, Dr. Glueck penned this momentous statement in a 1959 book:

> As a matter of fact, however, it may be stated categorically that no archaeological discovery has ever controverted a Biblical reference. Scores of archaeological findings have been made which confirm in clear outline or in exact detail historical statements in the Bible.[705]

Since Glueck was Jewish, it has been assumed that his use of the term "Biblical reference" refers only to the Old Testament. However, his research certainly included the New Testament, and this statement was made in a book the final chapter of which focuses on archaeological discoveries pertaining to Christianity.[706 707] Nevertheless, in the list above, the vast majority of items pertain to the Old Testament. This is because it covers a much longer expanse of time than the New, and the narrative often deals with the rise and fall of kingdoms, which are major sources of archaeological remains. In contrast, the New Testament is mostly concerned with Jesus and the early Christian Church.

I've heard it said faith is something you either have or you don't. I disagree. In many cases, faith is something that develops as observations warrant. There are certain people I trust implicitly. This is a result of many experiences in which I have witnessed them operate with honesty and integrity. I obviously haven't observed all of their actions, nor can I read their minds, but I have seen enough to have great faith in them. The same applies to the Bible. Although we do not have enough evidence to test every assertion in it, from what we can observe, the Bible's track record of credibility warrants a high level of faith in its contents.

CHAPTER 4

CONTINUITY

Near the end of the movie, *The Matrix*, a scene occurs in which the main character, Neo, is thrown against the wall of a subway. Parts of the wall shatter, and Neo along with fragments of the wall, fall to the ground between two railroad tracks. Yet, 45 seconds later, the same area between the tracks is shown, and it is inexplicably clear of the wreckage.[708] This is referred to as a discontinuity, and it is an axiom that discontinuities exist in all fictional works, no matter how brilliantly crafted. In the words of a famous legal scholar:

> [I]t is not possible for the wit of man to invent a story, which, if closely compared with the actual occurrences of the same time and place, may not be shown to be false.[709]

In my experience, I have found this to be true. Even the best fiction contains plot holes or discontinuities. We have already established in previous chapters that the Bible contains substantial amounts of accurate and verifiable information. In fact, while consulting research materials for this book, I was sometimes surprised to see non-Christian sources treating Biblical statements as historical facts.[710] What about, however, parts of the Bible for which we have no historical, scientific, or archaeological evidence to utilize as points of comparison? Is there any way to evaluate the credibility of these passages?

Yes. An excellent way to test any source for accuracy is to analyze it for continuity. The Bible is an ideal candidate for this manner of investigation because it is a compilation of 66 books written by numerous authors over the course of hundreds of years.[711] Such a diverse array of interrelated works offers multitudinous opportunities to contrast and compare.

For an even more stringent test of a work's reliability, one can collectively test all primary historical sources (Biblical and non-Biblical) that bear upon a matter for continuity. When we combine data from multiple sources, however, it is easy to forget what we are doing and fall into the trap of circular reasoning. To avoid this, we need to bear in mind that the objective of such investigations is to examine all available sources about a given matter and compare them to determine if a consistent story emerges.

Let's be clear, however, that we are not using the Bible to prove the Bible; we are testing the Bible to see if it agrees with itself and other independent sources—just as in a court case, in which witnesses recount their stories and physical evidence is examined. If everything adds up, this reflects well upon the witnesses' credibility. If unjustifiable inconsistencies emerge, odds are someone has a faulty memory or is not telling the truth.

We'll start the chapter with some matters regarding continuity within the Bible and culminate with a complex and demanding test of Biblical continuity involving a wide array of primary sources.

For the sake of not repeating the same phrases over and over again, I am not going to preface every assertion from the Bible with phrases such as, "The Bible claims…." For this chapter, this qualification applies universally. Thus, when you see a term like "the events of the crucifixion," what I really mean is "the events of the crucifixion as described in the Bible."

"What If God Was One of Us?"

We will start by addressing what some have declared to be a glaring contradiction in the Bible. The Book of John asserts that God loves us, and the Book of Jeremiah asserts that God has the power to do anything he wants.[712] How can these two claims be anything other than totally inconsistent with the reality that horrible things sometimes happen to people?

Think of someone you care for deeply and ask yourself, "If this person had cancer and I had the ability to cure him or her, would I?" Of course—that's a stupid question. So if God cares for us, why wouldn't He do the same? The dilemma posed by this question is at the root of some people's view that God cannot exist.

This viewpoint may appear to be logical, but it suffers from short-sightedness, for it fails to account for the scope and scale of eternity. If we are temporary beings, then death forever ends all that we are. Conversely, if our souls live for eternity after our physical bodies give way, then this life is an infinitely small fraction of our existence, and the trials we face and pleasures we enjoy will pale in significance to what follows.

Some have trouble grasping this idea because their thinking is restricted to the premise that there is no life beyond this one, which, as we saw in the chapter on science, is a very questionable assumption.

It's not that the opposite viewpoint is fully proven either, but the issue at hand is whether the Bible offers a logically consistent message on this subject, and there can be no doubt it does. Consider the following words of Jesus and observe how he places far more emphasis on the wellbeing of our souls than that of our bodies:

> And fear not them which kill the body, but are not able to kill the soul....[713]

> These things I have spoken unto you, that in me ye might have peace. In the world ye shall have tribulation: but be of good cheer; I have overcome the world.[714]

> And ye shall be betrayed both by parents, and brethren, and kinsfolks, and friends; and *some* of you shall they cause to be put to death. And ye shall be hated of all *men* for my name's sake. But there shall not an hair of your head perish. In your patience possess ye your souls.[715]

In this last passage, Christ tells his followers some of them would be put to death, and then two breaths later, says not a hair of their heads will perish. This is plainly not a contradiction but a blunt and effective way to draw a contrast between earthly and eternal life.

Many books, articles, debates, and lectures have been devoted solely to this question of why a loving God would allow us to suffer. To address this matter comprehensively would require a long diversion from the topic of continuity, and thus, our discussion will be concise. Yet, something more needs to be said before we move on.

Although the words of Jesus quoted above show he was focused on eternity, it is also vital to note he was not indifferent to our pain and suffering in this life. In fact, he cried after the death of his friend Lazarus. Yet, judging from the context of this event, it doesn't seem Jesus cried for Lazarus' sake because he was already planning to raise Lazarus from the dead and soon did so. Instead, it appears Jesus wept out of sympathy for Lazarus' sister Mary, who was heartbroken over her brother's death.[716] This brings to mind a popular song from 1995 called "One of Us." I'll let the lyrics speak for themselves:

> What if God was one of Us?
> Just a slob like one of Us?
> Just a stranger on a bus,
> trying to make His way Home

Tryin' to make his way Home
Back up to Heaven all alone
Nobody callin' on the phone
'Cept for the Pope maybe in Rome[717]

The songwriter appears to be oblivious to one of the most important points in the Bible: God *was* one of us.[718] He stepped off His throne to feel the pain of hunger, cruelty, betrayal, abandonment, and death.[719] He willingly subjected himself to these ordeals not to participate in a human reality show but to tear down the barriers that each of us has placed between ourselves and God (more about this in Chapter 10).

If there is an endless and tangible reality that can only be seen after this life has ended, how can we possibly place the events of this life in perspective from where we sit now? Pain, whether it be physical or emotional, is relative. A great number of things that elicit tears of anguish from a child will barely phase an adult. This is because the adult has a broader perspective. Does it break the heart of a mother and father to see their child weep over a lost toy? Absolutely, but Mom and Dad know this is not the end of the world, and it's going to be alright in the long run. Likewise, it hurts God to see us suffer, but his perspective is infinite, and he sees what we do not: eternity. Accordingly, the Bible cuts to the crux of this matter in the following passages:

> So we fix our eyes not on what is seen, but on what is unseen. For what is seen is temporary, but what is unseen is eternal.[720]

> For I reckon that the sufferings of this present time are not worthy to be compared with the glory which shall be revealed in us.[721]

Given everything above, it would be irrational to claim that the reality of suffering is inconsistent with what the Bible says about God's power and love for us.

More than 100,000 Connections between Biblical Passages

In evaluating the continuity of the Bible, we find an enormous amount of information to work with. A tool often used to research the Bible is known as the *Thompson Chain Reference*, and it specifies more than 100,000 links between interrelated Bible passages. To examine just a few of these instances, observe the interesting correlations between the following Old and New Testament passages.

Old Testament (Events well prior to the birth of Jesus)	New Testament (Events from the birth of Jesus onward)
Isaiah 53:7: He was oppressed, and he was afflicted, yet he opened not his mouth: he is brought as a lamb to the slaughter, and as a sheep before her shearers is dumb, so he openeth not his mouth.	**Mark 15:3–5:** And the chief priests accused him of many things: but he answered nothing. And Pilate asked him again, saying, Answerest thou nothing? behold how many things they witness against thee. But Jesus yet answered nothing; so that Pilate marveled.
Psalm 22:18: They part my garments among them, and cast lots upon my vesture.	**Mark 15:24:** And when they had crucified him, they parted his garments, casting lots upon them, what every man should take.
Zechariah 9:9: Rejoice greatly, O daughter of Zion; shout, O daughter of Jerusalem: behold, thy King cometh unto thee: he is just, and having salvation; lowly, and riding upon an ass, and upon a colt the foal of an ass.	**Matthew 21:5:** And the disciples went, and did as Jesus commanded them, And brought the ass, and the colt, and put on them their clothes, and they set him thereon. And a very great multitude spread their garments in the way; others cut down branches from the trees, and strawed them in the way. And the multitudes that went before, and that followed, cried, saying, Hosanna to the Son of David: Blessed is he that cometh in the name of the Lord; Hosanna in the highest.
Psalm 22:7–8: All they that see me laugh me to scorn: they shoot out the lip, they shake the head, saying, He trusted on the LORD that he would deliver him: let him deliver him, seeing he delighted in him.	**Matthew 27:41–43:** Likewise also the chief priests mocking him, with the scribes and elders, said … He trusted in God; let him deliver him now, if he will have him: for he said, I am the Son of God.
Isaiah 53:5: But he was wounded for our transgressions, he was bruised for our iniquities….	**1 Peter 2:24:** [Christ bore] our sins in his own body on the tree, that we, being dead to sins, should live unto righteousness….

Christians view the New Testament passages above (and many others) as the realization of Old Testament prophecies. On the other hand,
one could argue the New Testament writers simply mimicked the Old Testament. So, although similarities in the Bible can be very instructive and
the sheer number of them is incredible, we will see that dissimilarities are
what really enable us to test the Bible for continuity. Consider the following
example and its implications.

How Can Three Days and Three Nights Be Equivalent to Fewer Than 48 Hours?

All four Gospels assert Jesus was executed on a Friday afternoon[722]
and rose that Sunday morning.[723] Yet, in the Book of Mark, Jesus claims
he will rise on the "third day."[724] Likewise, in the Book of Matthew, Jesus
asserts he will "be three days and three nights in the heart of the earth."[725]
This appears to be a very concrete discrepancy. From Friday afternoon to
Sunday morning is fewer than 48 hours; certainly not three days and three
nights. However, a passage in an ancient Jewish work called the Jerusalem
Talmud explains that in the vernacular of ancient Israel, any "part" of a day
or a night was considered "a day and a night":

> A day and a night constitute a span, and part of a span is equivalent
> to the whole of it.[726]

Similarly, the first-century Jewish author Philo explains that with
regard to spans of time, the word "from" has two meanings in that it can
either include or exclude the starting point.[727] How many days are there
"from" Friday to Sunday? That depends upon whether "from" is used in an
exclusive or inclusive sense.

Understanding all of the above, when we examine the events of the
crucifixion, we see that Jesus was deceased during portions of Friday, Saturday, and Sunday.[728] Thus, applying the ancient Jewish definition of a "day
and night," it is perfectly consistent to say Jesus "rose on the third day" and
was "three days and three nights in the heart of the earth."

I find it amazing these historical records were preserved, for if they
were not, it would be logical to conclude that the Books of Matthew and
Mark conflict with themselves and each other. As it turns out, we have been
left with compelling evidence that the Gospels are thoroughly consistent
on this matter. Moreover, the subtle nature of this consistency makes it
almost certain it was not contrived. Apparent discrepancies such as this
that prove to be in agreement when viewed in the light of comprehensive

historical data are compelling evidence of a source's trustworthiness.

A Renowned Legal Scholar Dissects the Gospels

In the mid-nineteenth century, a Harvard professor by the name of Simon Greenleaf wrote a book in which he subjected the four Gospels to an examination from a legal perspective. Before we explore his findings, let's take a brief look at the author's credentials. According to the *Dictionary of American Biography*, Greenleaf was one of two people responsible for "the rise of Harvard Law School to its eminent position among the legal schools of the United States." He was described as "practical," "systematic," and "meticulously exact." His landmark three volume work, *A Treatise on the Law of Evidence*, came to be recognized as the "foremost American authority" on the matter of legal evidence.[729]

Although many gifted intellectuals of past and present have used their academic expertise to analyze the Bible, Dr. Greenleaf's mastery of legal evidence and his application of it have yielded some penetrating insights highly relevant to the matter of Biblical continuity. His incisive approach involves evaluating the continuity of the Gospels by subjecting them to the same tests used to judge the credibility of courtroom testimony.[730] [731] His conclusion is that the Gospels show no signs of being "creations of fiction" and every indication of being "true narratives of facts."[732] On what grounds does he stake these bold claims?

One of the key points recurring throughout his work is that the Gospels speak of the same basic events but are not exact copies of one another. Greenleaf notes that when multiple witnesses tell precisely the same story, we can be sure they conspired beforehand to ensure their stories coincide. In contrast, although the Gospels contain much common material,[733] there are many divergences that negate the possibility of dishonest collaboration.[734] In scrutinizing the details of these divergences, Greenleaf finds overwhelming evidence for the credibility and reliability of the Gospels.

For example, all four Gospels speak of Jesus performing a miracle in which he multiplies five loaves of bread and two fish into enough food to feed 5,000 men. The Books of Matthew and Mark say nothing about the location except that it was isolated and there was grass. The Book of Luke says nothing about the grass but states the event took place in a remote area of a town called Bethsaida. The Book of John identifies the location as a mountainside near the Sea of Galilee and notes the presence of grass.[735] It is obvious that at least three of the four accounts exhibit a measure of independence, yet watch the manner in which they tie together.

The Book of John says that before Jesus carried out the miracle, he asked a person named Philip where bread could be purchased. On its own, the fact that Jesus asked this question of Philip seems trivial. However, earlier, the Book of John states that Philip was from Bethsaida.[736] Recall from above that the Book of Luke places the event in this same town. Greenleaf notes that if you view these details in isolation, they seem "unimportant" and "gratuitous," but together, each renders the others:

> intelligible and significant. Jesus, intending to furnish bread for the multitude by a miracle, first asked Philip, who belonged to the city and was perfectly acquainted with the neighborhood, whether bread could be procured there. His answer amounts to saying it was not possible. These slight circumstances, thus collected together, constitute very cogent evidence of the veracity of the narrative, and evince the reality of the miracle itself.[737]

The points above constitute what are referred to as "undesigned coincidences." These occur when interlocking details substantiate each other in ways that were clearly not intended by authors and witnesses. They are a hallmark of authenticity and are found throughout the Gospels.[738] Over the course of 400-plus pages, Greenleaf provides numerous examples in which legal principles used to test the integrity of court testimony reveal "internal marks of truth" in the Gospels. These build upon each other until the cumulative weight of evidence becomes overwhelming in support of the conclusion that the Gospels are accurate accounts of real events.[739]

This brings us to a matter in which numerous interlocking ancient works come together to produce a remarkable testimony to the Bible's accuracy. This matter will constitute the vast bulk of this chapter, and although the question that follows is only a single question, the answer involves many facts bearing upon the issue of Biblical continuity.

WAS JESUS CRUCIFIED ON THE DAY THAT LAMBS WERE SLAIN FOR THE PASSOVER FEAST OR THE DAY AFTERWARDS?

In Chapter 1, we saw that the Book of John places the crucifixion of Jesus on the same day the Passover lambs were slain,[740] while in contrast, the Books of Matthew, Mark, and Luke place the crucifixion on the day after the Passover lambs were slain.[741 742 743] This apparent contradiction has been an issue of debate for many centuries. In fact, there is record of a dispute relating to this matter going back more than 1,800 years.[744 745]

In modern times, various theories have been proposed to explain this seeming discontinuity, but after much reading, I found none convincing. Of course, the simplest explanation is that one or more of the Gospels is wrong, but given the forthcoming information, such a conclusion is entirely irrational.

Through gathering and comparing primary sources of data on the calendars of ancient Israel, I was led to evidence that firmly resolves the matter. Furthermore, this evidence reveals several interesting undesigned coincidences that speak to the Bible's veracity. Note that if someone else has previously arrived at any part of what is detailed below and I have not credited him or her, please accept my apologies. So much has been written about the Bible that many lifetimes would be needed to read it all.

If the genuine answer to the issue of the crucifixion day has eluded detection for more than 1,800 years, one might presume this is not a simple matter, and such is the case. Herein lies a great challenge, for even James Clerk Maxwell, the renowned physicist whose incredible scientific achievements were detailed in Chapter 2, acknowledged, "I know the tendency of the human mind to do anything rather than think." In this harried age in which we live, authors strive to produce materials that are quickly and easily comprehended, but how does one handle a subject matter that is inherently complex?

In such instances, writers often choose to distill their work down to the point where readers are simply forced to take their word for it. I could follow suit, but this would defeat the purpose of this book, which is to bring forward enough concrete evidence to allow readers to develop their own rational conclusions. Although this process can require intense thought, in the same speech quoted above, Maxwell declared that even though thinking is "a process from which the mind naturally recoils," when it is:

> completed, the mind feels a power and an enjoyment which make it think little in the future of the pain and throes which accompany the passage of the mind from one stage of development to another.[746]

Hopefully, your effort applied to what follows will reward you with the kind of power and enjoyment Maxwell extolled. With that, here is a basic overview.

During the first century A.D., most people of Jewish faith followed a sect known as the Pharisees. Their calendar coincides with the account of the crucifixion in the Book of John and was used to determine the date of the crucifixion in Chapter 1. However, other groups preferred different calendars, and the Books of Matthew, Mark, and Luke coincide with one of these. When the Gospels are placed side by side with the Israelite calendars, all the data fits together seamlessly to yield two major conclusions:

1) Jesus celebrated Passover one day before the Pharisees did.

2) The one day variance between the Book of John and the other Gospels is a result of the fact that the Book of John utilizes a different calendar from the other Gospels.

As we will see, extensive historical evidence points to this conclusion, and the circumstances defy the possibility of coincidence or fabrication. Furthermore, this historical evidence independently corroborates the Gospels in several other respects, adding even more force to their credibility. This represents a prime example of what Greenleaf considered to be persuasive evidence for the continuity and accuracy of the Gospels.

Now let's move on to the comprehensive explanation. Recall from Chapter 1 that Jesus lived in the land of Israel and was Jewish.[747] [748] In about 93 A.D., Josephus identified three major sects of Judaism that had been in existence for a "great while." These were the Essenes, Sadducees, and Pharisees.[749] [750] This, by the way, corroborates the Bible in that the Pharisees and Sadducees are mentioned in the Gospels.[751] We will soon see that each of these three major sects had their own calendar.

Some have claimed Jesus was an Essene, but both Josephus and Philo describe the Essenes, and there are some glaring contradictions between their religious rules and the actions of Jesus.[752] [753] [754] [755] It is also clear that Jesus was not a Pharisee or a Sadducee, for he criticized both sects and highlighted specific issues that he had with each of them.[756] In summary, we cannot assign a calendar to Jesus based upon his being a member of one of these sects. Of course, this doesn't rule out the possibility he used the same calendar as one of them.

To put this in context, one must realize how important calendars were to the people of ancient Israel. In our society, most of us don't know or care that we use a Gregorian calendar.[757] In contrast, these ancient peoples considered it a religious duty to celebrate holy days on the proper days and struggled to determine exactly when these days were.[758] As we will see, they did not take this lightly, and it was a source of great contention. Some sects placed so much emphasis on their calendar that they made it acceptable to violate the rules of their holy days so tasks related to the calendar could be carried out properly. For example, carrying a burden on a holy day was forbidden, but this rule could be ignored if you were traveling to perform a calendar-related task.[759] (I surmise they considered it preferable that a few people broke the rules of a holy day rather than risking the entire community celebrating a holy day on the wrong date.) With this in mind, let's compare the events of the crucifixion with the calendar information that survives from this era.

DID JESUS OR THE GOSPEL WRITERS USE THE CALENDAR OF THE DEAD SEA SCROLLS?

Let's start with a simple matter. Starting in the 1940s and into the 1950s, a mass of ancient Jewish documents dating from the 3rd century B.C. to the 1st century A.D. were discovered near a ruins called Khirbet Qumran.[760] [761] These are the famous Dead Sea Scrolls. Among them are 11 documents containing details about an ancient solar calendar.[762] Some scholars think these scrolls were owned by a community of Essenes, but there is legitimate debate regarding whether or not this was the case.[763] [764] There is no need for us to take a long detour and sort this out because in either instance, the calendar is Jewish and dates to in or around the first century. Therefore, it is relevant to the issue at hand.

That being said, as we attempt to test the Gospels for continuity with each other and with separate sources of this era, it is easy to rule out the possibility that any of the Gospel writers used this solar calendar. In the calendar of the Dead Sea Scrolls, the Passover celebration always took place on a Wednesday.[765] There is no way to reconcile this with the Bible because all four Gospels state that Jesus was crucified on a Friday.[766] Once again, the Books of Matthew, Mark, and Luke state that Jesus celebrated Passover the evening before he was crucified,[767] [768] [769] while the Book of John says Passover was celebrated in the evening immediately after Jesus was crucified.[770] There is no rational way to make the jump from either of these scenarios to a Passover celebration on a Wednesday. Thus, we can be certain the Gospel writers did not use the calendar of the Dead Sea Scrolls.

THE UNDESIGNED COINCIDENCES BEGIN TO MOUNT

How about the calendar of the Pharisees? Did any of the Gospel writers use their calendar? As detailed in Chapter 1, the vast majority of people followed the Pharisees, and thus, we used their calendar to calculate the date of the crucifixion. The fact that the Book of John coincides with their calendar and a lunar eclipse confirms the date is compelling evidence that the Book of John employs this calendar or one that is indistinguishable from it.

But how about the other Gospels? Either they are wrong or Jesus and his disciples did not abide by the Pharisees' calendar and these books followed suit. If the latter is the case, why would the Book of John use a different calendar from Jesus? As will be shown in the present and next chapter, the Book of John was probably the last Gospel written, and it was penned at a time when every sect of Judaism except for the Pharisees had been practically eliminated.[771] [772] Given this, it is understandable why John would use the calendar of the Pharisees for a frame of reference. It is important to note, however, that although the Book of John concurs with the calendar of the Pharisees, it does not assert or even vaguely imply that Jesus and his disciples used this calendar.[773]

Here is the place where it is crucial to pay close attention. In the Jewish religion, the Passover feast is a holy day.[774] In accordance with this, the Books of Matthew and Mark assert that when the "chief priests, elders, and scribes" assembled to plot against Jesus, they specifically stated they did not want him arrested and executed on the same day as the Passover feast because they feared an "uproar of the people."[775] This is substantiated by the writings of Josephus, which confirm that an execution on a holy day would have been cause for public outrage.[776] This point is critical because the Books of Matthew and Mark, along with the Book of Luke, lay out the events of the crucifixion in this manner:

a) Jesus ate the Passover feast with his disciples in the evening.

b) Later that night, he was arrested.

c) In the morning, Jesus was brought before Pontius Pilate and underwent a trial.

d) Later that day, he was crucified.[777] [778] [779]

From the perspective of our modern calendar (in which each new day begins and ends at midnight), these events take place over the course of two calendar days. However, in the calendar of the Pharisees, this is not the case. In their calendar, the dividing line between days was the "evening twilight."[780] This means each new day began around sunset, which is also the case with the modern Jewish calendar and Jewish calendars in general.[781] [782] [783] Knowing this and reexamining the events of the crucifixion as described by Matthew, Mark, and Luke (see chart on next page), we see that by any Jewish calendar, all of the events (Passover feast of Jesus and his disciples, arrest, trial, crucifixion) occur on precisely the same calendar day.

Midnight	Sunset	Midnight	Sunset	Midnight

- - - - Friday in Jewish Calenders - - - -

Passover Feast | Arrest | Trial | Crucifixion

- - - Thursday in Our Calender - - - | - - - - Friday in Our Calender - - - -

Remember, the writings of Josephus corroborate the Books of Matthew and Mark in that an execution on a holy day would have caused an "uproar of the people." Again, who were the people? As Josephus recorded, "the body of the people" performed their "Divine worship, prayers, and sacrifices" according to the "directions" of the Pharisees.[784] Thus, it was their Passover feast that was to be avoided. Yet, the same Gospels (Matthew and Mark) that state the religious leaders ruled out executing Jesus on the day of the Passover feast also affirm Jesus ate the Passover feast on the very same day he was executed. In other words, when we understand how Jewish calendars function, we see that Matthew and Mark clearly imply Jesus did not celebrate Passover on the same day as the Pharisees. This is a great example of an undesigned coincidence, and the implications will be apparent as we move forward.

THE OTHER CALENDAR

If Jesus didn't follow the calendar of the Dead Sea Scrolls or that of the Pharisees, is there any other he and the Gospel writers might have used? Much historical data points to such a calendar, but we must correlate this data before its workings become evident. After this is done, we will see this calendar meshes perfectly with the accounts of the crucifixion in Matthew, Mark, and Luke—and moreover, there are good reasons to believe these books use it. As detailed in the approaching academic text, this calendar was based upon a monthly occurrence called a lunar conjunction and generally ran one or two days ahead of the Pharisees' calendar. In addition to Jesus and his disciples, a faction of the Sadducees called the Boethusians, a people known as the Samaritans, and a colony of Jews who lived on an island in the middle of the Nile River all used this calendar at one time or another.

For those who will brave the forthcoming academic text, my advice is to not get so enmeshed in the details that you lose sight of the big picture. The next six pages establish the existence of an ancient calendar and detail its workings. This is a complicated subject, and thus, discussion of it requires much space. While you are in this space, however, remember this is not just about calendars. The primary matter before us is to determine whether or not the collective data of the Gospels and the other writings of this era are in agreement.

▶At this point, it will be helpful to brush up on some material from Chapter 1 and add a little background. In the first chapter, citations 198–212 detail how calendar practices found in several ancient Jewish works (the Mishnah, Tosefta, Babylonian Talmud, and Jerusalem Talmud) were used by the Pharisees. Additionally, the appendix (page 321) shows that the doctrines found in the Mishnah generally mirror those of the Pharisees and explains that the other three works consist of excerpts from the Mishnah surrounded by interpretations and supplemental information. Also recall that Josephus was a Pharisee. We are about to quote extensively from the writings just mentioned. Other than the review just provided, I will not detail how each calendar practice cited below from these works is related to the Pharisees. Enough evidence has been provided already to show that the ancient calendar rules found in them generally concur with those of the Pharisees.

Boethusian Rhapsody

Three of the four ancient Jewish works mentioned above describe one or more calendar disputes with a group called the Boethusians. There is not enough historical data to be absolutely certain of their roots, but the widespread scholarly opinion is that they were Sadducees who were descendants of a priest named Boethus.[785] [786] The primary source of information on Boethus is Josephus, who identified three of his sons and a grandson who were appointed high priests between 24 B.C. and 44 A.D.[787] The Encyclopaedia Judaica asserts that the Boethusians existed for a 100-year period between 30 B.C. and 70 A.D.[788] Both of these relatively short spans of time turn out to be centered around the earthly life of Jesus, which was from about 2 B.C. to 33 A.D.[789] With further regard to the timeframe in which the Boethusians flourished, the Babylonian Talmud quotes a preeminent Jewish teacher in the middle of the first century A.D.

as engaging the Boethusians in a calendar dispute.[790][791]

We learn from Josephus, the Babylonian Talmud, and the Tosefta that the Boethusians/Sadducees held positions of authority but were forced to follow the dictates of the Pharisees.[792][793][794] Because of this, it has been assumed they used the calendar of the Pharisees even though they took issue with it. However, we don't know if this was always the case, and Josephus identifies a period in the first century B.C. when the Sadducees clearly had the upper hand.[795] Whatever the situation may have been, the key fact to grasp is that the Boethusians took issue with the calendar of the Pharisees.

In the calendar used by the Pharisees, the first day of each month started when the crescent of the new moon became visible.[796] However, the new moon is not always visible to everyone at the same time. As the *Jewish Encyclopedia* explains:

> At first, a small and faint arc, like a sickle can be seen by those endowed with good sight, from spots favorable for such an observation. It may therefore happen that in different places the reappearance of the moon is noticed on different days.[797]

In order to make sure each month began at the very first appearance of the new moon, a system was established whereby the general public kept an eye out for it. Those who claimed to have seen the moon went to a courtyard in Jerusalem where they were interrogated to determine if they were telling the truth.[798] However, a problem arose with the system. As the Mishnah explains:

> Beforetime they used to admit evidence about the new moon from any man, but after the evil doings of the heretics they enacted that evidence should be admitted from them that they knew.[799]

In expounding upon this passage, the Babylonian Talmud tells us the Boethusians paid two people to say they had seen the new moon when in fact they had not. The plot was exposed, and the authorities who oversaw the calendar decided they would only accept testimony about the new moon from people they knew.[800][801]

To understand what the Boethusians were up to requires another refresher on the calendar of the Pharisees. The Pharisees correctly determined that the length of a lunar cycle was about 29½ days. In keeping with this, they required that all months be either 29 or 30 days long. To ensure no month could be shorter than 29 days, any testimony that would cause this to occur was automatically rejected. Likewise, to ensure no month could be longer than 30 days, if the new moon were not spotted by the 30th day of a month, a new month would automatically begin on the next day.[802] [803] This means there was only one way to trick the Pharisees with regard to the new moon: If the new moon did not appear on the 30th day, claim that you saw it. If they believed you, a new month would begin on that day, which means the outgoing month would be shortened to 29 days. The end result was that the new month would begin one day earlier than it was supposed to.

Now here we have something—a Jewish sect who coexisted with Jesus and attempted to manipulate the calendar of the Pharisees so the month would start one day early. This lines up with the chronology in the Books of Matthew, Mark, and Luke, all of which run exactly one day ahead of the Book of John. Of course, it is possible the Boethusians sabotaged the Pharisees' calendar simply out of spite. Or maybe the Boethusians preferred a different calendar, and since the general public and possibly the Boethusians themselves were bound by the rules of the Pharisees, they tried to trick the Pharisees into doing it their way.

Part of the reason this mystery has been so difficult to unravel is that the Boethusians/Sadducees were virtually wiped out when the Romans destroyed the Jewish Temple in 70 A.D., and none of their writings have survived. Therefore, we don't have any internal perspectives on their practices; only an occasional rant against them by one of their enemies.[804] [805] However, we are not stuck at this point because another group of people who lived in ancient Israel also tried to shift the calendar of the Pharisees ahead by one day.

THE BAD SAMARITANS

The name Samaritans was applied to a people who inhabited a region of Israel known as Samaria. As this map shows, to the south of Samaria was Judea, and to the north was Galilee,[806] which is where the Bible says Jesus lived.[807]

Living adjacent to the Israelites, the Samaritans shared some points of commonality with their neighbors, primarily in that the first five books of the Jewish Bible and the first five books of the Samaritan Bible are almost the same. Because of this, both the Israelites and the Samaritans celebrated the holy days, such as Passover, ordained in these books. Of course, dissimi-

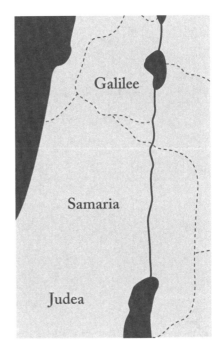

larities between the two groups existed as well. For example, the other 34 books of the Jewish Bible are not even a part of the Samaritan Bible.[808][809][810][811]

Both Josephus and the New Testament assert that the Jews and Samaritans did not get along (again, more corroboration for the Bible's accuracy).[812][813] Among their quarrels, Josephus records that some Samaritans entered the Jewish Temple on Passover and "threw about dead men's bodies."[814] Obviously, this is an act of desecration, making it unlikely that the Samaritans were celebrating a holy feast such as Passover on that day. This, of course, implies they used a different calendar. Josephus places this event in the same era that the Bible places the birth of Jesus.[815]

The Mishnah describes a communication system used by the Pharisees to inform people when a new month had begun. It consisted of a series of fires that were kindled on certain mountains.[816] A 29-day-month was signified by fires on the mountains, and a 30-day-month was signified by the absence of fires.[817]

According to the Mishnah, the Samaritans tried to sabotage this system.[818] Because of how the process worked, the only way the Samaritans could have done this was to light fires for an outgoing month that contained 30 days. This would trick the people into believing the outgoing month contained 29 days. The result would be the same as the fraud perpetrated by the Boethusians. The new month would begin one day earlier than it was supposed to.

As a result of the Samaritans' actions, the fire signals were done away with and replaced with human messengers.[819] With regard to timing, one scholar places this sabotage of the calendar by the Samaritans between 141 and 109 B.C.[820] Another places it in the era of a prominent rabbi who flourished near the middle of the first century A.D.[821][822] I am not confident in either of these conclusions, but it's pretty safe to say the event took place after 200 B.C.[823] and before 70 A.D.[824]

A little-known fact is that the Samaritans have managed to survive to this present day. As of February 2007, there were exactly 704 of them.[825][826] They have preserved some ancient history, but the oldest explanation of their calendar was penned in the 12th century or later.[827][828][829] This is a long way off from the first century, but we can move slightly closer to our period of interest because the critical element of this calendar as far as we are concerned is confirmed in a 10th-century work written by a member of a rival faith. The author criticizes the Samaritans for their practice of starting each new month based upon calculations used to determine when the sun, moon, and earth would line up, as shown on the next page.[830]

In this two-dimensional sketch duplicated from Chapter 1, assume the earth, moon, and sun are not in the same plane. If they were, it would

New Moon

Dark side of the moon
faces the earth

© iStockphoto.com/Eraxion/janrysavy/knickohr

result in a solar eclipse. What we see depicted here is referred to as a lunar conjunction. This occurs regularly at intervals of 29 days, 12 hours, 44 minutes, and 2.8 seconds.[831] (When the moon is lined up on the other side of the earth, the result is a full moon.)

When lunar conjunctions occur, the dark side of the moon directly faces the earth, and the moon becomes invisible to us. The Samaritans calculated when conjunctions were due to occur and set their calendar so that each conjunction triggered the start of a new month. The writer who criticized this practice favored starting each new month when the crescent of the new moon became visible,[832] which was the same system used by the Pharisees.[833] After conjunction, the new moon takes anywhere from 15–54 hours to appear.[834] Thus, a calendar based upon conjunctions would contain months that ran a day or two ahead of the Pharisees' calendar.

We know the Samaritans in the tenth century based their calendar on conjunctions, but how do we determine what they did in the first century?[835] Based upon the disputes detailed above, it is clear the Samaritans and Pharisees were not using the same calendar at that time. However, it may be that the Samaritans used observation of the new moon like the Pharisees, but the result was sometimes different because local weather and geography caused the Samaritans to spot the new moon on different days from the Pharisees. Some have assumed the Samaritans used this method because it "was used by almost all people" in this region of the world including ancient "Arabs and Jews."[836] This presumption, however, is not completely factual. For our period of interest, there is only evidence that the Pharisees preferred this method. As we have seen, the Boethusians, like the Samaritans, tried to undermine it, and the authors of the Dead Sea Scrolls used an entirely different system.

DISCOVERY FROM AN ISLAND IN THE NILE RIVER

An intriguing archaeological find made near the onset of the 20th century sheds some more light on this matter. On an island that sits in the middle of the Nile River in Southeast Egypt,[837] some ancient Jewish writings were unearthed.[838][839] The documents were penned on papyrus in the 5th century B.C. and are quite extraordinary.[840] In addition to their antiquity, some of the documents contain two dates: one from a calendar used by their Jewish authors and a corresponding date from the Egyptian calendar.[841]

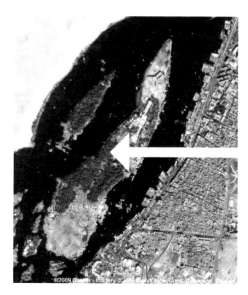

Elephantine Island, Aswan, Egypt

The Egyptian dates are based upon a well-understood solar calendar of 365 days per year, while the Jewish dates are based upon a lunar calendar. By using astronomy to find the intersection points for each set of dates, one can determine the date of each document in terms of our calendar.[842] When this is done, we find the Jewish dates are not based upon observations of the new moon but are generally consistent with a system based on lunar conjunctions.[843][844]

The scholar who first figured this out in the early 1900s was brilliant in that he did it from scratch and without the aid of computers. Yet, he made the common mistake of assuming all Jews who lived in Israel during the 1st century A.D. used observation of the new moon to trigger the start of each new month. He noted that the calendar used in these documents was more technically advanced than one based upon observations of the new moon, and since technology tends to progress over time, he supposed this calendar was never used in Israel or it would have become the dominant system. What he failed to account for were the reluctance of people to change their religious traditions and the calendar disputes detailed above.[845] ◄

INDEPENDENT EVIDENCE KEEPS POINTING TO THE SAME CONCLUSION

What we have is a distinct pattern among the opponents of the Pharisees. Whether we are dealing with the Samaritans, the Boethusians/Sadducees, or Jesus and his disciples, historical records unmistakably imply their calendars sometimes ran a day ahead of the Pharisees' calendar. When we isolate each of these groups, not enough information exists to state for a fact how each of their calendars operated, but when we look at the combined evidence, a coherent picture emerges. Furthermore, we have good reason to look at the evidence in just this way, for beyond the fact that these three groups all lived in Israel during the same era, they were also linked in significant ways—particularly with respect to their differences with the Pharisees.

In addition to their attempts to deceive the Pharisees into starting their months a day early, both the Samaritans and Boethusians celebrated a holy day called "the Sheaf" according to the same timing, which differed from that of the Pharisees.[846] Beyond this, an ancient Jewish work that reflects the doctrines of the Pharisees[847] groups the "daughters" of the Samaritans and Sadducees together and insinuates they are both unclean.[848] [849] [850]

The Bible states that the Pharisees accused Jesus of being "a Samaritan."[851] Perhaps this was meant as a general insult, or, more logically, maybe the Pharisees recognized similarities between Jesus and the Samaritans. Recall from above that Samaria was located between Judea and Jesus' homeland of Galilee. Josephus wrote, "It was the custom of the Galileans, when they came to the holy city at the festivals, to take their journeys through the country of the Samaritans."[852] The Bible tells us this is exactly what Jesus did.[853] Thus, Galileans and Samaritans may have shared commonalities simply due to geographic proximity.

Perhaps most importantly, the Pharisees observed a tremendous number of religious rules not found in the Bible. Both Jesus and the Sadducees took major issue with this. Josephus explains:

> [T]he Pharisees have delivered to the people a great many observances by succession from their fathers, which are not written in the laws of Moses [i.e. first five books of the Bible]; and for that reason it is that the Sadducees reject them, and say that we are to esteem those observances to be obligatory which are in the written word, but are not to observe what are derived from the tradition of our forefathers. And concerning these things it is that great disputes and differences have arisen among them....[854]

This corroborates the Gospel of Matthew, which states that Jesus rebuked the Pharisees for teaching doctrines not commanded in the Scriptures.[855] Clearly, this is the type of sweeping disagreement that would have manifested in many ways, which certainly could have included calendar differences. All of this tells us it is sensible to examine the evidence concerning the calendars of Jesus, the Sadducees, and the Samaritans collectively. And again, when we do this, we find a consistent pattern of their calendars running one day ahead of the Pharisees' calendar and a rational reason for the difference: the use of lunar conjunctions to trigger the start of new months.

THE DATA ADD UP PERFECTLY

The historical and archaeological evidence above converges upon the inescapable conclusion that a calendar (or calendars) based upon lunar conjunctions was used by various people living in Israel during the first century. So how do the Gospels line up with such a calendar? If we reconstruct it during the year and month that we calculated for the crucifixion, we find it runs exactly one day ahead of the Pharisees' calendar.[856] This is precisely the scenario implied by the collective evidence of the Bible and the ancient sources that enlighten the historical context of this event.

The significance of all this data interlocking in the manner it does is difficult to appreciate, so let's recap. The Book of John gives an account of the crucifixion that concurs with the calendar of the Pharisees, placing the execution of Jesus on the calendar day before the Passover feast. However, this book doesn't assert or even vaguely imply that Jesus used this calendar. The other three Gospels affirm that Jesus celebrated the Passover feast on the same calendar day he was executed, yet two of these Gospels state the religious leaders ruled out executing Jesus on the day of the Passover feast to avoid an "uproar of the people," thus implying Jesus did not celebrate Passover on the same day as "the people." Josephus says there were divisions among the people, but the majority followed the dictates of the Pharisees. Hence, the combined data above places the Passover feast of Jesus exactly one calendar day before the Passover feast of the Pharisees. Concurring with this, extensive historical and archaeological evidence reveals the existence of an ancient calendar used by opponents of the Pharisees that ran exactly one day ahead of the Pharisees' calendar in the same year and same month as the crucifixion. This can hardly be coincidental. It is extraordinary that so many independent pieces of data fit together in such a seamless and rational manner.

If this weren't enough, through investigating this matter, we found the Bible is further corroborated by other ancient writings on several significant and peripheral points including:

(1) The existence of Jewish sects called the Sadducees and (2) the Pharisees

(3) Carrying out the death sentence on a holy day would have enraged the populace of Jerusalem.

(4) The existence of the Samaritans and (5) their adversarial relationship with the Jews

(6) Galileans journeyed to Jerusalem to celebrate their holy festivals and (7) passed through Samaria on the way.

(8) The Pharisees abided by various extra-Biblical religious traditions.

Does Any Evidence Conflict?

Among the hundreds of pages of ancient historical writings I scrutinized, I found three quotes that appeared to diverge from the picture painted by the rest of the evidence. After researching them, however, I concluded that none cast doubt upon or even slightly weakened the implications of the rest of the evidence. Nevertheless, I encourage you to establish your own opinion by evaluating the data yourself.

1) An ancient Jewish work (the Mishnah) contains a passage indicating the death penalty was imposed "on" a holy feast. This doesn't appear to fit well with the Bible's assertion that the religious leaders wanted to avoid arresting and executing Jesus on the day of the Passover Feast. Below is the relevant passage:

> ▶He was not condemned to death either by the court that was in his own city or by the court that was in Jabneh [which sat there from A.D. 70 to 118], but he was brought up to the Great Court that was in Jerusalem. He was kept in guard until one of the [three] Feasts [Passover, Pentecost, or Tabernacles] and he was put to death on one of the [three] Feasts, for it is written, *And all the people shall hear and fear, and do no more presumptuously* [quote from the Old Testament, Deuteronomy 17:31]. So [says] R. Akiba. But R. Judah says: They should not delay his judgment but put him to death at once, and write out and send messengers to every place, [saying], 'Such-a-one the son of such-a-one has been condemned to death by the court'.[857]◀

This dispute involves two conflicting opinions concerning the death

penalty. In sum, three times a year, Jews traveled to Jerusalem to participate in holy feasts.[858] [859] One rabbi cited an instance in which a death-row
inmate was held until one of these feasts so the public would be present and his execution would serve as an example to potential offenders.
Another rabbi thought it best to carry out the execution immediately upon
conviction and dispatch messengers to convey news of the punishment to
the public.

Besides the fact that there was an obvious disagreement over this issue,
a mass of evidence from Josephus makes it unmistakably clear that an execution on a Sabbath (Saturday) or a feast day would have been cause for
outrage. The Pharisees and Jews in general considered Sabbaths and feasts
such as Passover to be holy days of rest in which almost all work was prohibited.[860] [861] [862] Josephus tells us that in the second century B.C., a thousand
Jewish men, women, and children were killed during an attack in which they
refused to fight back because:

> they were not willing to break in upon the honor they owed the
> sabbath, even in such distresses; for our law requires that we rest
> upon that day.

After this massacre, the Jews decided they would act defensively in
such situations.[863] [864] But even so, they continued to endanger their lives
for the purpose of observing the Sabbath. In the first century B.C., a battle
broke out between the Romans and some Jews. The Romans figured out
that Jews would not attack on the Sabbath, so they took advantage of this
and used these days to move their weaponry into strategic positions without fear of an offensive by the Jews.[865] If the Jews considered Sabbaths so
holy that they would imperil their own lives to observe them in a battle
situation, it is exceedingly doubtful they would carry out executions on
such days.

Furthermore, Josephus wrote it was forbidden even to cry on a festi-
val.[866] Again, not a great day for inflicting the death penalty. He also tells
us that Augustus Caesar (reigned from 31 B.C. to 14 A.D.) and one of his
generals ordered that Jews were not required to appear before judges on
the Sabbath.[867] In short, Josephus confirms the Bible by making it plain
that an execution on a holy day would have triggered an "uproar."

So is this passage in the Mishnah indicating the death penalty was
carried out "on" one of the feasts inaccurate? I don't think so. I just think
it needs to be clarified for modern readers. For the audience to whom it
was written, it was entirely obvious that executions were not to take place

on holy days. There was no need to spell this out for them. They would automatically understand that the phrase "on one of the [three] Feasts" did not refer to the calendar day of the feasts but to the general occasion of the feasts—in other words, while the people were in town for the feasts. The idea was to make a public example of the criminal, not to desecrate a holy day.

▶2) The Tosefta lists a series of regulations pertaining to when it is permitted to use "leaven belonging to Samaritans."[868] After these rules are given, the following is stated:

> Under what circumstances [have the forgoing specifications been laid down]? When [the Samaritans] did not prepare their Passovers along with the Israelites [on the same day], or if they pushed the holiday up by one day.

This fits perfectly with all of the evidence above, but the passage continues:

> But if they prepared their Passovers along with the Israelites' [celebration], or if they pushed the holiday back by one after the Israelites' [celebration], leaven belonging to them is permitted immediately after the end of Passover.[869]

The problem here is that it is not possible for a calendar based on conjunctions to run behind a calendar based upon observations of the new moon. This is because the absolute earliest that the new moon becomes visible is at least 15 hours after conjunction.[870] Given that this figure is under 24 hours, both calendars can synchronize in certain months, but a conjunction-based calendar cannot run behind an observation-based one. Thus, at first glance, the last part of the sentence quoted above implies that the Samaritans did not use a calendar based upon conjunctions.

However, this regulation was in all likelihood listed to cover a hypothetical situation. The incredibly detailed and hairsplitting nature of the Tosefta makes it very plausible this regulation was only included for the sake of completeness.[871] To paraphrase the Tosefta: "These are the rules if the Samaritans celebrate Passover before us, on the same day as us, or after us." Given this, it would be unreasonable to presume this passage undermines the other evidence indicating the Samaritans used a calendar based upon conjunctions.

3) The Tosefta offers a motivation for the Boethusians' attempt to trick the Pharisees with regard to the new moon; namely, the Boethusians' desire to celebrate a holy day called Pentecost in accordance with their timing:

At first they [the sages] would accept testimony concerning the new moon from everybody. One time the Boethusians hired two witnesses to come and fool the sages. For the Boethusians do not concede that Pentecost should come at any time except on the day following the Sabbath.[872]

If the attribution of this motive is correct, it wouldn't contradict the deduction that the Boethusians preferred the use of conjunctions to trigger the start of new months, but it would weaken the case for it by explaining their attempted sabotage of the Pharisees' calendar in a different context.

That being said, the *Jewish Encyclopedia* doesn't accept this motive, and rightfully so.[873] The Old Testament commands that Pentecost be observed 50 days after a celebration called the Sheaf.[874] [875] As stated above, the Boethusians and Pharisees had differing rules for the timing of the Sheaf, but by either of their rules, it took place in the same month as Passover.[876] This means that if the Boethusians wanted to trick the Pharisees so that Pentecost would fall according to their reckoning, they would have to fool them into starting the month of Nisan on a different day, which would also cause Passover to take place on a different day. In other words, to get the timing of Pentecost right in their eyes, the Boethusians would also have to spoil the timing of their Passover. There is no logic in this unless the Boethusians also wanted to celebrate Passover according to a different timing from the Pharisees, which brings us right back to the Boethusians using conjunctions.◄

ACCURATE AND INDEPENDENT WITNESSES

There is more corroborating evidence we could examine, but there is no need to do so. The point has already been sufficiently made that this is a stunning example of what Simon Greenleaf found in his examination of the Gospels: an apparent discrepancy that is resolved in the light of information provided by numerous independent testimonies, all coming together to produce a coherent picture. Such circumstances amount to exceptionally strong evidence that the Bible contains accurate and independent witnesses to the same events.[877]

HOUR OF THE CRUCIFIXION

Are there any inconsistencies in the Bible to which there are no known solutions? Absolutely. Consider, for example, that the Book of Mark states Jesus was crucified at the "third hour," while the Book of John states Jesus was crucified around the "sixth hour."[878] The general understanding of this convention is that the first hour refers approximately to the time between 6 and 7 A.M. Thus, the third hour would be around 9–10 A.M. and the sixth hour from about 12–1 P.M.[879] Does this show the Bible is contradictory and therefore false in at least some respects?

Far from it. In probing the Gospels for continuity on matters such as the length of time Jesus was in the tomb and the day of the crucifixion, we have access to critical historical evidence that allows us to see a broader picture. This also allows us to conduct the most thorough test of a work's continuity. In this case, however, such evidence may well be lost among the multitudes of ancient writings that are no longer in existence. As explained in Chapter 1, there is no telling exactly how many works have been lost over time, but because ancient historians referenced the writings of their contemporaries, we are aware of many works that once existed but are now lost.[880] [881] Clearly, given the calendrical mayhem that prevailed in ancient Israel, it would be irrational to exclude the possibility that this seeming discrepancy regarding the hour of the crucifixion stems from different conventions for labeling the hours of a day. In addition to everything we have learned above, consider what the Roman scientist Pliny the Elder wrote in 77 A.D.:

> The days have been computed by different people in different ways. The Babylonians reckoned from one sunrise to the next; the Athenians from one sunset to the next; the Umbrians from noon to noon; the multitude, universally, from light to darkness; the Roman priests and those who presided over the civil day, also the Egyptians and Hipparchus, from midnight to midnight.[882]

Furthermore, this discrepancy may be the result of a copying error, which as we will see in the next chapter, is a distinct possibility. In Greek, the language of the New Testament, the number 3 is denoted by the letter Γ (gamma) and the number 6 by the letter Ϝ (digamma).[883] The visual similarity between these letters makes such an error quite possible.

YOUR TURN IN THE COURTROOM

In the course of this chapter, we examined a small portion of what is a massive pool of evidence, focusing primarily on the four Gospels because they are foundational to Christianity. Whether or not these books agree with each other in every particular instance is not provable or disprovable. We simply don't have enough historical data to make such a claim. However, one thing is certain: the Gospels relate to one another in some very distinctive ways that overwhelmingly support the credibility of their testimonies. This is reinforced to an even greater degree when we consider other independent sources from this era.

To sum up and paraphrase Greenleaf's logic,[884] suppose you are called to jury duty and are presented with this case: A home was broken into and robbed. The individual who robbed the house used a knife to break into the window, and while he was forcing the window open, the tip of the blade broke off. That night, the police apprehended a person running down an alleyway. The man had a knife on him, and the tip of it was broken off. It matched the piece of blade that was found at the home. He also had in his possession the jewelry that was taken from the home, and his fingerprints matched those on the windowsill of the house.

If you were on this jury, short of strong contradictory evidence, would you vote to convict this person? I would, because the evidence demonstrates beyond a reasonable doubt that he is guilty. Except in cases in which a video is available, jurors don't actually see people commit crimes. Instead, they rely on evidence, and if the evidence is substantive and credible, they apply it to make a decision. If one can use evidence of this nature to send someone to jail, why not also use it to establish or solidify one's faith?

CHAPTER 5

ANCIENT MANUSCRIPTS, MODERN BIBLES, AND *THE DA VINCI CODE*

Although we have already learned a great deal about how academic disciplines interrelate with and corroborate Biblical texts, we have gotten a little ahead of ourselves. Given that the Bible is a collection of books passed down to us through a process of hand copying over many ages, how do we know that the text we read today accurately reflects what the original authors wrote thousands of years ago? For example, in the King James Version of the Bible, which is one of the most popular Bible translations, the Book of First Kings states:

And Solomon had <u>forty</u> thousand stalls of horses for his chariots, and twelve thousand horsemen.[885]

Yet, the Book of Second Chronicles states:

And Solomon had <u>four</u> thousand stalls for horses and chariots, and twelve thousand horsemen....[886]

So which is it—"forty thousand stalls" or "four thousand stalls"? In all probability, this is the result of a copying error. In Chapter 1, we discussed various difficulties ancient copyists had to contend with such as the absence of spaces between adjoining words. As a result of this and other challenging circumstances, we noted "errors were common" in the copying process. These mistakes, however, tend to be insignificant. Additionally, numerals are the type of text most prone to such errors because when a number is copied improperly, it is frequently not evident from the context that a mistake has been made.[887]

In some ancient manuscripts, both of the verses cited above contain a figure of four thousand, and Bible translations such as the New International Version reflect this.[888] We will be discussing the issue of different Bible versions in more depth, but for now, one obvious point should be recognized. It is quite conceivable that someone could read "four thousand" and inadvertently write "forty thousand" or vice versa. In Hebrew, the word "four" is changed into "forty" when two letters are added.

four .... ארבע

forty ...... ארבעימ

People like me who consider the Bible to be inspired by God might object to the conclusion that this was a copying error. For if this is the case, how can we know that any of the Bible has been passed down to us accurately? Moreover, how does anyone know that copyists did not deliberately alter the texts that comprise the Bible in the intervening thousands of years since they were written? The answer, as will be shown, is that there is a tremendous amount of concurring evidence for the authenticity of the Biblical texts. Before we examine this, however, it behooves us to learn about the nature of such evidence.

Evaluating the Authenticity and Age of Ancient Manuscripts

There are many factors to consider when trying to ascertain how reliably an ancient work has been transmitted to modern times. One such factor is the amount of time between when a work was written and the oldest copies in existence. As a rule of thumb, the shorter the span, the less chance there is for inaccuracies to creep into the text. In other words, a copy of a copy has more chance of being accurate than a copy of a copy of a copy of a copy. However, because of the human element, this is not always the case. A lone careless copyist can produce a flawed manuscript just as surely as eight meticulous copyists can pass down a text from one to the next with incredible accuracy.

This begs the questions of how we determine when a work was written and when the copies were made.

One of the surest ways to determine when a work was written is through explicit chronological information recorded by the author. For instance, in the last paragraph of the *Antiquities of the Jews*, Josephus writes:

> [U]p to the present day, which belongs the thirteenth year of the reign of Domitian Caesar....[889]

Domitian became emperor in 81 A.D.,[890] which means the thirteenth year of his reign was approximately 93 A.D. Hence, in all probability, Josephus completed this work at about this time. In the absence of strong contradictory evidence, scholars take such chronological information at face value, but this obviously leads to yet another question: How do we know Domitian became emperor in 81 A.D.? Since Domitian was ruler of the Roman Empire, there are multiple interlocking historical works and archaeological finds attesting to the date of his reign. Suffice it to say that because of this, we can be almost certain in identifying timeframes for the reigns of most Roman emperors.

In many cases, authors of ancient works did not record such explicit chronological details, but we can often establish the latest possible date of a work by a reference to it from another work that is datable. Scholars also examine the events, idioms, and thematic elements of a writing to see if they are indicative of a certain era. This is not always precise or concrete, but in most cases, enough combined evidence emerges to get us somewhere in the ballpark.

How then, do scholars go about dating the manuscripts of these ancient literary works? Such manuscripts rarely contain a date indicating when they were copied. However, many ancient legal and business documents, such as wills and memorandums, do contain dates. By analyzing these, paleographers (handwriting specialists) have determined that different letter styles, formats, and abbreviations prevailed during different time periods. Using a framework derived from this information, paleographers estimate when undated manuscripts were copied. This, again, is not an exact science, but it is generally considered to be the best method available, and some believe it to be accurate within a range of about 50 years for our era of interest.[891] [892]

In evaluating how reliably a work has been transmitted, one of the best forms of evidence we can have to work with is an ancient translation of the work in question.[893] This is because when a book was translated into another language, it created detached streams of manuscripts that can be used to check one another. For instance, if a work was written in Latin in 100 A.D. and translated into Greek in 300 A.D., we can be fairly confident that where the manuscripts in both languages agree with one another, the text has been accurately transmitted since 300 A.D. To a certain extent, the same applies in cases wherein ancient writers quoted from other ancient writers. In a nutshell, what we are looking for is independent corroboration—the earlier, the better.

The Old Testament

Let's see how the Old Testament stacks up according to these criteria. With the exception of 21 sentences written in an ancient language called Aramaic, the Old Testament was composed in Hebrew.[894] (For the sake of simplicity I will hereafter refer to these Hebrew/Aramaic manuscripts as "Hebrew.") The various ages of the 39 books in the Old Testament is a complicated matter of much debate that we are not going to settle here, but accounting for a diverse range of scholarly opinions, it is fairly certain that the oldest books were written no earlier than 1,450 B.C. and the newest no later than 100 B.C.[895] [896] For the record, I am of the opinion that the books of the Old Testament were written between the 15th and 5th centuries B.C.

How many Old Testament manuscripts exist and when were they copied? Just to start with, there are more than 6,000 Hebrew manuscripts dating from the 9th century A.D. up through the invention of the printing press in the 15th century.[897] [898] [899] Moreover, the relatively recent discovery of the Dead Sea Scrolls unveiled manuscripts dating to the era of Jesus and including portions of every Old Testament book except Esther.[900] [901] Man-

uscripts of this age tend to be severely decayed, and it is rare to find a book that is totally intact.[902] However, in the 1940s, a manuscript of the entire book of Isaiah copied in about 100 B.C. was discovered, making it 1,000 years older than any other previously known Hebrew manuscript of this book.[903] [904] [905]

Furthermore, there exists a Samaritan Bible of uncertain vintage with manuscripts dating from the Middle Ages. This is also written in Hebrew and parallels the first five books of the Jewish Bible.[906] [907]

© iStockphoto.com/BryanLever

Cave where some of the Dead Sea Scrolls were found

With respect to translations, the Old Testament was translated into Greek by Egyptian scholars in the third and second centuries B.C.[908] [909] There exists a manuscript of this entire translation that dates to the 4th century A.D. and various older manuscripts containing portions of it.[910] [911] After the Greek translation, the Old Testament was also translated into Latin, Aramaic, Arabic, Syriac, and Ethiopic.[912]

Put simply, the quantity, antiquity, and translational evidence for the Old Testament manuscripts is exceptional. For the purpose of comparison, consider the following. The Latin and Greek classics were written between about 700 B.C. and 250 A.D.[913] [914] By and large, ancient translations of these works are non-existent, and the manuscripts we know them through were copied in the 9th–15th centuries A.D.[915] [916] For example:

WORK	AUTHOR	ERA WRITTEN	NUMBER OF MANU- SCRIPTS	EARLIEST MANU- SCRIPT
History of the Peloponnesian War[917]	Thucydidies	5th century B.C.	7	11th century A.D.
The Gallic War[918]	Julius Caesar	1st century B.C.	6	9th century A.D.
The Histories[919]	Polybius	1st century B.C.	3	10th century A.D.
The Lives of the Twelve Caesars[920]	Suetonius	2nd century A.D.	150+	9th century A.D.

The enormous number of Old Testament manuscripts makes comparing and contrasting them a formidable task. In the main, they agree with one another, but there are some notable divergences we will examine. As will be shown, there are three main textual sources for the Old Testament: the traditional Hebrew text, the Greek text, and the Samaritan text. The Dead Sea Scrolls corroborate all of these in certain respects, but lend far more credence to the traditional text.

▶The 6,000+ Hebrew manuscripts mentioned above (dating from the 9th through the 15th centuries A.D.) were copied by scribes who frequently used clever and effective methods to ensure the writings were transmitted accurately. For instance, on each line of text, the letters were counted and compared with the source document to make certain the number was the same. Furthermore, the letter and word located at the midpoint of each section of text was identified and checked against the source document. As a result of such exhaustive efforts, these manuscripts agree with one another to a remarkable extent. The text of these manuscripts is referred to as the traditional or "Masoretic" text.[921][922] (The word Masoretic stems from the Jewish word meaning "tradition."[923]) This text is the basis of the modern Hebrew Bible and of the Old Testament in most Christian Bibles.[924] [925] To gain an understanding of how minuscule the variants are among these manuscripts, examine this divergence in Proverbs 15:20:

> A wise son makes a glad father, but a foolish <u>man</u> despises his mother.

> A wise son makes a glad father, but a foolish <u>son</u> despises his mother.[926]

Actually, this is one of the larger deviations I have seen. Most involve minor details, such as spelling.◄

One of the primary facts revealed by the Dead Sea Scrolls is that the traditional text of the Old Testament has been transmitted very accurately for at least two millennia. The amazing correlation between some of the Dead Sea Scrolls and the traditional text provides concrete proof that this text has not materially changed since the time of Jesus.[927][928][929] Other Dead Sea Scrolls diverge from the traditional text, and we'll discuss them in a moment, but regardless of any other evidence, it can be stated for a fact that the traditional text we hold in our hands today is essentially the same as what ancient Hebrews read 2,000 years ago. Moreover, given that the oldest Dead Sea Scroll dates to the third century B.C. and matches the traditional text almost perfectly,[930][931] we can easily stretch this figure to more than 2,200 years in certain cases.

▶Regarding the Dead Sea Scrolls that diverge from the traditional text, some of them display similarities to the Samaritan Bible and others to the ancient Greek translation of the Hebrew Bible.[932] (For simplification, I will hereafter refer to these as the "Samaritan" and "Greek" texts, and use the word "Scrolls" for the Dead Sea Scrolls.) Although these particular Scrolls don't mirror the Samaritan and Greek texts to the same degree that other Scrolls match the traditional text,[933][934] they reveal that the Samaritan and Greek texts are also of ancient origin.

This is where the matter gets thorny, for although the traditional, Samaritan, and Greek texts basically coincide with each other,[935][936] they do contain some obvious differences. In such cases, how is it decided which is closer to the original text? The traditional text is generally considered to be the most accurate representation of the original writings,[937] but many scholars think that in certain passages other texts reflect the original writings better. For example, in Genesis 4:8 the traditional text reads:

> Cain said to his brother Abel. And when they were in the field......

In contrast, the Samaritan text reads:

> Cain said to his brother Abel: "Let us go to the field." And when they were in the field....[938]

The context seems to require the "extra" words included in the Samaritan text. Thus, some scholars attribute the "missing" words in the traditional text to a copying error. Bible translations such as the New International Version use the traditional text as a foundation but rely upon other texts in certain places such as the instance above.[939] [940] Other translations such as the New King James Version rely almost exclusively on the traditional text and contain footnotes indicating where significant alternative readings exist.[941]

Some scholars think equal weight should be given to all of the Biblical texts and that each divergence between them needs to be evaluated on an individual basis. I take issue with this, however, because in examining the logic behind such decisions, I see a process fraught with subjectivity. This is illustrated by a few quotations from a book I have cited heavily, *Textual Criticism of the Hebrew Bible*. The author, Emmanuel Tov, is one who takes the view that each text should be given equal weight.[942] Though I disagree with this and other conclusions in his book, his command of the facts is nothing short of amazing, and I have learned a great deal from him. Tov is very candid about the uncertainty inherent in choosing between alternative readings:

> This procedure is as subjective as subjective can be.[943]

> By means of a comparison of texts it is possible to identify deliberate changes, but the decision on what exactly comprises such a change necessarily remains subjective.[944]

> Furthermore, it must be realized that even if there are objective aspects to the rules [of textual evaluation], the very selection of a particular rule remains subjective.[945]

Along the same lines, Tov makes the statements below regarding scholars employing a process called emendation, which is using the context of a passage to derive readings that are not found in any manuscripts:

The procedure of emending the biblical text is one of the most subjective aspects of textual criticism in particular, and of biblical research in general. Generally speaking, in the course of the past few centuries, far too many emendations have been suggested, most of which may now be considered unnecessary.[946]

Difficulties arise particularly in the areas of language, vocabulary and the exact meaning of the context. ... A reasonable amount of self-criticism is also required with regard to the limits of our knowledge, especially in the area of language. Due to the fact that the available data in this area is very fragmentary, it may be that an apparently incorrect or unsuitable reading was, nevertheless, the original one.[947]

I consider this last quote generally applicable to all types of textual evaluation. What may seem from the context to be a doubtful reading could simply be the result of the original author writing in a non-typical manner. Furthermore, alternative text that seems plausible may be the result of a copyist "fixing" what lay before him because it sounded strange. In which case, the reading that seems less plausible may in reality be the original one.

Instead of the endless speculation often associated with textual evaluation, let's deal with the facts we know for certain. We know for a fact that the traditional text has been faithfully transmitted since the time of Christ. Contrast this with the Greek text, in which the manuscripts contain a "maze of manifold variants." These discrepancies often make it difficult to determine exactly what the Greek text even says. In fact, scholars sometimes use the term "Old Greek" to signify what each believes, in his or her "judgment," to be the original text.[948] [949] [950] [951] Likewise, the Samaritan text was not copied with great consistency.[952] In contrast, we have undeniable proof that the traditional text was transmitted with the utmost care through the extensive religious, cultural, and political upheavals that took place between 250 B.C. and 1500 A.D. How this could not be credited as evidence for its general trustworthiness is beyond me.◄

From even prior to the time of the earliest Biblical manuscripts, there is still other evidence that supports the authenticity of the traditional text. One simple illustration involves a foreign ruler named Sargon, who was an Assyrian king in the 8th century B.C.[953] When archaeologists uncovered an inscription dating from this era bearing Sargon's name, the spelling matched the traditional text perfectly, but in the Greek text, the spelling somehow became "Arna." This is just one example of what is typically the case.[954]

I'm not asserting that only the traditional text should be relied upon and everything else ignored, just that its track record of reliability should make it authoritative unless there is clear and objective evidence to the contrary. The other texts undoubtedly have value, however, as evidenced by the fact that the New Testament was composed entirely in Greek and contains many quotes from the Greek translation of the Old Testament.[955][956]

MODERN BIBLES

All this being said, I think people sometimes make far too much of minor differences between manuscripts. When a Biblical book is translated, someone must decide which manuscript or manuscripts are going to be used as the basis of the translation. Given the fact that the Bible is the holy book of many millions of people, imagine how a dispute regarding which manuscripts to use could escalate.[957]

Some ascribe letter for letter perfection to certain Bible versions and claim all others are heretical. When I read such viewpoints, I am reminded of the manner in which Jesus castigated the Pharisees for being so concerned with the miniscule details of Scripture that they missed the important points. In addition to describing the Pharisees as "blind guides," Jesus told them their actions were the equivalent of picking a tiny bug out of their food while at the same time swallowing an entire camel.[958] In accord with Jesus' message of keeping the big picture in perspective, he summarized hundreds of pages in the Old Testament with two basic points:

> Jesus said unto him, Thou shalt love the Lord thy God with all thy heart, and with all thy soul, and with all thy mind. This is the first and great commandment. And the second is like unto it, Thou shalt love thy neighbor as thyself. On these two commandments hang all the law and the prophets.[959]

▶Some will point to the following passage when making the case that only a certain Bible version can be the trustworthy:

> [Jesus said:] Think not that I am come to destroy the law, or the prophets: I am not come to destroy, but to fulfil. For verily I say unto you, Till heaven and earth pass, one jot or one tittle shall in no wise pass from the law, till all be fulfilled.[960]

Since the words "jot" and "tittle" refer to the smallest characters in the Hebrew alphabet,[961][962] some interpret this passage to mean that Jesus said the Scriptures will be preserved so perfectly that not a single letter will ever change. Given the context, I think a more logical interpretation is that Jesus used the phrase "jot and tittle" as a rhetorical tool to drive home the point that every aspect of the Scriptures would be fulfilled. However, that's just my opinion, so let's dig a little deeper.

The person who takes the view that Jesus was saying no letter of Scripture will ever change has a huge contradiction to contend with. Consider this: Various books of the Bible quote passages from other books of the Bible, and often there are differences. Here is an example from the King James Version, which is the Bible I read most often. In the Book of Matthew, a single chapter before Jesus speaks of the "jot and tittle," he quotes three times from the Old Testament book of Deuteronomy. In the first two instances, the wording in the New Testament is almost exactly the same as in the Old Testament,[963] [964] but not in the last:

> Matthew 4:10: "Then saith Jesus unto him ... it is written, Thou shalt worship the Lord thy God, and him only shalt thou serve."

> Deuteronomy 6:13 "Thou shalt fear the LORD thy God, and serve him...."

I am not aware of any collection of Biblical manuscripts in which all of the books display perfect agreement in wording.[965] [966] Thus, anyone who attaches excessive theological importance to such matters is at odds with the manuscripts that underlie whatever Bible they may read.◄

This is not to say every Bible version is a good one. Translators can easily undermine the Scriptures by using baseless and irrational word choices to support their personal agendas. I am familiar enough with the King James Version, New King James Version, and the New International Version to feel very comfortable in their reliability. This doesn't mean I think all others are unreliable, only that I don't know enough about them to make such a determination.

By far, the largest divergence among the various Old Testament texts pertains to the fact that the Greek text contains several chapters and books that are not a part of the traditional text. The Bibles of the Catholic, Greek Orthodox, and Russian Orthodox Churches contain these chapters and some of these books, whereas Protestant and Jewish Bibles do not.[967] [968] This accounts for some of the theological differences between various denominations. For example, Catholics believe in a place called purgatory, where the souls of people who have died are cleansed of their sins prior to entering heaven.[969] This is based primarily upon a passage in a book called 2 Maccabees, which is part of the Catholic Bible but not the Protestant.[970]

In short, the churches that include these chapters and books in their Bible primarily cite their traditions for accepting them.[971] Protestants reject these books because the Jewish Bible does not contain them, and the New

Testament quotes from the Old Testament 239 times but never from any of these books.[972] [973] [974] [975] [976] Thus, the argument is made that they were not considered Scripture by the writers of the New Testament. In contrast, the New Testament of almost all Christian churches—including the Catholic, Protestant, Greek Orthodox, and Russian Orthodox Churches—contains exactly the same books.[977] We will come back to this idea of how and why certain books came to be included in the Bible in a moment.

THE NEW TESTAMENT

To put it mildly, the evidence concerning New Testament manuscripts is compelling. Taking into account a broad range of scholarly opinions, we can safely say the earliest books of the New Testament were written after 48 A.D. and the latest before 150 A.D.[978] [979] Personally, I don't think there is enough evidence to confidently date every book, but my best estimate is that they were all written in the period from about 50 to 120 A.D.

There exist more than 5,000 New Testaments manuscripts written in their original tongue, which is Greek.[980] The oldest is a fragment from the Book of John that dates to between 100 and 125 A.D. The handwriting indicates it may have been copied in the late first century, but because such a date brings us so close to when scholars think the Book of John was written, some scholars find it difficult to accept this.[981] [982] A full two-thirds of the New Testament text is found in manuscripts that date prior to 300 A.D.[983]

Furthermore, early Christian writers from the second century onward quoted from the books of the New Testament so profusely that even if there weren't a single New Testament manuscript in existence, the vast majority of it could be reconstructed from the writings of these early Christians.[984] [985] Also, prior to 600 A.D., various New Testament books were translated into Latin, Armenian, Arabic, Persian, Nubian, Ethiopic, Gothic, Syriac, and Coptic.[986] [987] In addition to the multitude of manuscripts in these languages, there are 8,000 New Testament manuscripts written in Latin.[988]

In summary, the evidence for the textual accuracy of the New Testament books is overwhelming. With the exception of about two paragraphs in the entire New Testament, the manuscript evidence is so strong there is no rational basis for any kind of uncertainty over the substance of the text.[989] [990] [991] [992] [993]

WHY DID CERTAIN BOOKS BECOME PART OF THE BIBLE?

There can be little doubt that the New Testament books have been passed down to us accurately, but a tougher matter to determine is how and why each of these books came to be included in the Bible. Much historical evidence exists on this topic but no totally conclusive answer. Historical records from the second century onward contain lists made by various Christians of the writings they considered to be holy. These generally include the four Gospels and the letters of Paul. However, not until 367 A.D. do we find a list of New Testament Scriptures that is unequivocally and exactly the same as our modern Bible. Up through this period, it is clear that no central authority dictated what these books would be. As best as can be seen from the historical record, the New Testament took its shape through the consensus of early Christian churches.[994] [995] Some objective criteria for inclusion have been proposed, such as whether or not the author personally knew Jesus or one of his apostles, but there isn't enough evidence to know for certain who authored each and every one of the 27 books in the New Testament.[996]

Simply put, for Christians, it is a matter of faith that God worked through early Christians to guide the Bible to its present form.[997] What is not a matter of faith, however, is that these books provide many remarkable indications of reliability and accuracy. As we have already seen, this evidence extends into academic disciplines such as history, earth science, physiology, archaeology, and neurobiology. Additionally, in the upcoming pages, many more disciplines will be shown to support the Bible.

THE DA VINCI CODE

While we're on the subject of how the books of the Bible came to be, this is an appropriate time to address *The Da Vinci Code*. Since this novel sat atop every major best-seller list in the U.S. and was made into a movie starring Tom Hanks,[998] it hardly needs introduction. However, very briefly, *Publishers Weekly* describes it as "an exhaustively researched page-turner about secret religious societies, ancient coverups and savage vengeance." It's sad that we even need to discuss a novel in this venue, but the author's claims coupled with the gullibility of many people to believe them make it advisable to do so. It is also important to realize that this is but one example from a recent throng of popular fiction and pseudo-documentaries that masquerade as scholarship.

The author of *The Da Vinci Code*, Dan Brown, begins the book with a page entitled "FACT." Listed under this word are four sentences including the following: "All descriptions of artwork, architecture, documents, and secret rituals in this novel are accurate." So although this is a work of fiction, the author emphasizes that the book has a firm foundation in reality. This notion is reinforced by book reviews such as one in the *Chicago Tribune* stating that the author transmits "several doctorates' worth of fascinating history and learned speculation."[999]

We're now going to take a look at some passages from the book to get an idea of how unfounded this claim of "FACT" truly is. What we will witness goes well beyond the regrettable and all-too-common practice of making assertions that have no verifiable basis. It's one thing to make a claim without evidence to support it, but quite another do so in the face of definitive evidence to the contrary.

To follow the context of the quotes below, realize they come from the mouths of characters named Leigh Teabing, a "former British Royal historian" who was "knighted by the Queen"; Sophie Neveu, an "attractive" 32-year-old French police agent, and Robert Langdon, a "Professor of Religious Symbology" at Harvard University.

Church on "Sunday"

The Da Vinci Code, pages 232–233:

> "Originally," Langdon said, "Christianity honored the Jewish Sabbath of Saturday, but [the Roman Emperor] Constantine shifted it to coincide with the pagan's veneration day of the sun." He paused grinning, "To this day, most churchgoers attend services on Sunday morning with no idea that they are there on account of the pagan sun god's weekly tribute—*Sun*day."

To understand just how ridiculous this is, realize that Constantine was born in about 285–290 A.D.[1000] Yet, no later than 161 A.D,[1001] a Christian by the name of Justin Martyr wrote the following:

> But Sunday is the day on which we all hold our common assembly, because it is the first day on which God, having wrought a change in the darkness and matter, made the world; and Jesus Christ our Savior on the same day rose from the dead.[1002]

The historical record is clear. Christians were attending church on Sunday for at least 120 years before Constantine was even born. This has been the practice of Christians for as long as anyone knows, and the quote above is only one of many that prove Brown wrong.[1003] [1004]

To be comprehensive, it should be noted there are some small elements of truth mixed into Brown's words. The word "Sunday" derives its name from the word "sun," and Constantine made Sunday a day of rest in the Roman Empire. However, he did this in 321 A.D., more than 150 years after Justin Martyr wrote the above. Additionally, I am unaware of any historical source that states or implies this was done "to coincide with the pagan's veneration day of the sun." In direct opposition to this, a work published 17 years after Constantine issued his decree specifically states its purpose was to free people of their work so they could attend church services.[1005] [1006]

DID EARLY FOLLOWERS OF JESUS CONSIDER HIM TO BE THE SON OF GOD?

The Da Vinci Code, page 233:

> [Teabing] "…Constantine needed to strengthen the new Christian tradition, and held a famous ecumenical gathering known as the Council of Nicaea. … [U]ntil *that* moment in history, Jesus was viewed by his followers as a mortal prophet—a great and powerful man, but a *man* nonetheless. A mortal."
>
> [Sophie] "Not the Son of God?"
>
> "Right," Teabing said. "Jesus' establishment as the 'Son of God' was officially proposed and voted on by the Council of Nicaea."

Again, this is demonstrably false. The Council of Nicaea took place in 325 A.D.[1007] Well before this, the historical record is replete with quotes from followers of Jesus who considered him to be the Son of God. No later than 161 A.D.,[1008] Justin Martyr wrote:

> In these books, then, of the prophets we found Jesus our Christ foretold as coming, born of a virgin, growing up to man's estate, and healing every disease and every sickness, and raising the dead, and being hated, and unrecognized, and crucified, and dying, and rising again, and ascending into heaven, and being called, the Son of God.[1009]

If that weren't enough, in about 111 A.D. (200+ years before the Council of Nicaea), a Roman statesman known as Pliny the Younger wrote a letter to the Emperor of the Roman Empire in which he described his procedure for identifying and punishing Christians. He declared that Christianity was a "cult" and former Christians had informed him that they met on a regular basis and sang "in honor of Christ as if to a god."[1010]

I cannot emphasize enough that these are just two of the many sources that puncture Brown's claim. Also, let's not overlook the four New Testament Gospels. Regardless of whether or not you consider these to be accurate in every detail, it is undeniable they were written at least two centuries before the Council of Nicaea. Their exact ages are uncertain, but the Encyclopædia Britannica dates them as follows:

Mark: 64–70 A.D.
Matthew: 70–80 A.D.
Luke: about 80 A.D.
John: "late-first-century"[1011]

Give or take 20 years from these dates, and you have a range that encompasses practically all educated opinions.[1012] As mentioned earlier, there exists a manuscript containing a fragment of John that dates from 100–125 A.D. There is also a manuscript containing portions of Matthew that dates from 125–150 A.D., one of Luke from 125–150 A.D., and one of Mark from about 200 A.D.[1013] Note that in 29 places, the four Gospels refer to Jesus as the "Son of God." Thus, to claim that Jesus was not viewed by his followers as the son of God until 325 A.D. is beyond absurd.

Hijacking Jesus' Human Message to Expand Their Power?

The Da Vinci Code, page 233:

> [Teabing] "Many scholars claim that the early Church literally *stole* Jesus from His original followers, hijacking his human message, shrouding it in an impenetrable cloak of divinity, and using it to expand their own power."

Recall from the first chapter that in about 115 A.D., the Roman historian Tacitus detailed the following events that took place in 64 A.D.:[1014]

> Nero fastened the guilt and inflicted the most exquisite tortures on a class hated for their abominations, called Christians…. Mockery of every sort was added to their deaths. Covered with the skins of beasts, they were torn by dogs and perished, or were nailed to crosses, or were doomed to the flames and burnt, to serve as a nightly illumination, when daylight had expired.[1015]

If the early Church was looking to "expand their own power," they sure picked an inconvenient way to go about it. In the context of these events, Tacitus described Christianity as a "most mischievous superstition."[1016]

Clearly, these people were not tortured and killed for a "human message." Take special note that these events happened about 30 years after Jesus was crucified and 260 years before the Council of Nicaea.

DID CONSTANTINE SELECT THE GOSPELS OF THE NEW TESTAMENT?

The Da Vinci Code, page 231:

> [Teabing] "More than *eighty* gospels were considered for the New Testament, and yet only a relative few were chosen for inclusion—Matthew, Mark, Luke, and John among them."
> "Who chose which gospels to include?" Sophie asked.
> "Aha!" Teabing burst with enthusiasm. "The fundamental irony of Christianity! The Bible, as we know it today, was collated by the pagan emperor Constantine the Great."

Once again, this is ridiculous. A man by the name of Origen Adamantius was a very learned and prolific Christian writer who lived from about 185 to 254 A.D.[1017] In other words, he died before Constantine was even born. In a work entitled *Commentary on Matthew*, Origen named the four Gospels that appear in our modern New Testament (Matthew, Mark, Luke, and John) and wrote that these "alone" are "uncontroverted in the Church of God under heaven."[1018] And like before, this is just one of several historical records that contradict Brown's claim.[1019]

Were there other "gospels" in circulation? Of course. What was to stop anyone from picking up a pen and writing a story about Jesus? People still do this in the present day. The relevant point is not the existence of other so-called gospels but whether or not they show any indication of historical value, which is what we'll address next. Also, note that I was unable to find even a shred of substantiation for the claim that there are as many as 80 such gospels, and Dan Brown has certainly never provided it.

ARE THE DEAD SEA AND NAG HAMMADI SCROLLS THE EARLIEST CHRISTIAN DOCUMENTS?

The Da Vinci Code, pages 233, 244–245:

> [Teabing] "The Dead Sea Scrolls were found in the 1950s hidden in a cave near Qumran in the Judean desert. And of course, the Coptic Scrolls hidden in 1945 at Nag Hammadi. In addition

to telling the true Grail story, these documents speak of Christ's ministry in very human terms." …

"These are photocopies of the Nag Hammadi and Dead Sea scrolls, which I mentioned earlier," Teabing said. "The earliest Christian records. Troublingly, they do not match up with the gospels in the Bible."

Let's deal with a simple matter first. As we have already seen, many of the Dead Sea Scrolls are manuscripts of Old Testament books, which turn out to prove that these works have been passed down to us with tremendous accuracy. Other Scrolls contain information such as the calendar we discussed in the last chapter and writings with certain theological parallels to the New Testament.[1020] However, none of them even mention Jesus. Some people have sought to identify anonymous figures in the Scrolls with Jesus, his brother James, John the Baptist, and the Apostle Paul, but as a scholar who specializes in the Scrolls has explained, such theories fall into the category of "improbable speculations."[1021] [1022] Ironically, after mentioning the Dead Sea Scrolls, *The Da Vinci Code* never quotes from or even appeals to them. In brief, the Scrolls tell us nothing of Jesus and play no role in the plot of *The Da Vinci Code*.

As a brief aside to correct some of Brown's tangential misinformation, it should be noted that the scrolls from Qumran were not found in "a" cave, but in 11 caves; and not in the "1950s," but from 1946/47 though 1956.[1023] Also note that several ancient manuscripts found in the Judean desert at locations other than Qumran are often classified as Dead Sea Scrolls.

Returning to the main issue, let's discuss the texts Brown actually appeals to, which are two texts, found at Nag Hammadi, referred to as the Gospel of Philip and the Gospel of Mary Magdalene.[1024]

The so-called Gospel of Philip is about 20 pages long, and I emphasize the phrase "so-called," because this text doesn't even attempt to narrate the events of Jesus' life.[1025] [1026] It mainly consists of theological musings with 15 sayings of Jesus woven into the text. The earliest and only manuscript we have of it dates to about 350 A.D.,[1027] and the work itself was written somewhere around 200 A.D.[1028] [1029] Obviously, it is not one of the "earliest Christian records." It postdates the New Testament Gospels by roughly 100–130 years, and half of the sayings of Jesus that appear in it are found in the New Testament Gospels.[1030] Moreover, in two of these cases, the Gospel of Philip employs a common introductory phrase for citing Scripture: "The word says…."[1031] In other words, the Gospel of Philip is dependent upon the New Testament Gospels and cites them as authorities.[1032]

Like the Gospel of Philip, the Gospel of Mary Magdalene is not a narrative of Jesus' life. It consists of dialogues among Jesus and his disciples, all of which take place after Jesus has risen from the dead.[1033] [1034] So much for the claim that the Nag Hammadi Scrolls "speak of Christ's ministry in very human terms." The oldest manuscript of this work dates to the beginning of the third century, and the range of scholarly opinions regarding when it was written stretches from the early to late second century.[1035] This 100-year range of speculation exists because the text is extremely short and lacking in chronological details. About half the text is missing, and what remains fills only about three pages.[1036]

Philip, Mary, and other "gospels" that are not in the New Testament are often referred to as "apocrypha" or "apocryphal gospels." One common thread that runs through such works is the tendency to fill in blank areas left by the New Testament. For example, after Jesus' childhood, his father Joseph disappears from the pages of the Bible without explanation. What became of him? Did he play any other role in Jesus' life? The Gospel of Philip allegedly enlightens us on this matter:

> Joseph the carpenter planted a garden because he needed wood for his trade. It was he who made the cross from the trees which he planted. His own offspring hung on that which he planted. His offspring was Jesus and the planting was the cross.[1037]

According to the New Testament, there was a 40-day period between when Jesus rose from the dead and when he ascended into heaven. The New Testament says that during this time, he spoke of "things pertaining to the kingdom of God."[1038] However, very few details are provided. But fret not—there is no need to wonder. In addition to the Gospel of Mary, there are at least eight other works among the Nag Hammadi Scrolls that claim to convey what Jesus said after he rose from the dead.[1039]

When we set the New Testament and apocryphal gospels side by side, the pattern becomes clear. Where the New Testament leaves an empty space, these works jump in to fill it. Such writings about Jesus were composed throughout ancient times and all the way up through the Middle Ages. They are simply a manifestation of the natural human desire to "know more."[1040] [1041] Again, what's to prevent anyone from picking up a pen and writing a story about Jesus?

Was Jesus Married to Mary Magdalene?

I considered omitting this next section because it is somewhat detailed and pertains to a quote from the Gospel of Philip, which bears no signs of historical reliability. However, since it is central to the plot of *The Da Vinci Code*, I thought it worthwhile to address.

The Da Vinci Code, pages 245–246:

> [Teabing] "As I said earlier, the marriage of Jesus and Mary Magdalene is part of the historical record." …
>
> Flipping toward the middle of the book, Teabing pointed to a passage. "The Gospel of Philip is always a good place to start."
>
> Sophie read the passage:
>
> *And the companion of the Saviour is Mary Magdalene. Christ loved her more than all the disciples and used to kiss her often on her mouth. The rest of the disciples were offended by it and expressed disapproval. They said to him, "Why do you love her more than all of us?"*
>
> The words surprised Sophie, and yet they hardly seemed conclusive. "It says nothing of marriage."
>
> *"Au contraire."* Teabing smiled, pointing to the first line. "As any Aramaic scholar will tell you, the word *companion*, in those days, literally meant *spouse*."

I'm not sure if the word "companion" meant "spouse" in Aramaic and I'm not going to take the time to find out because the Gospel of Philip wasn't written in Aramaic. It was written in Greek. And the only existing manuscript is a translation into a language called Coptic. There are some words in the Gospel of Philip associated with a language similar to Aramaic called Syriac, but "companion" is not one of them. In this work, "companion" is translated from the Coptic word *koinwnoc*, which comes from the Greek word κοινωνός (or *koinōnos*).[1042] [1043] [1044]

This word can have any number of meanings, including marriage partner, business partner, partner in faith, partner in suffering, partner in adultery, etc.[1045] [1046] In cases like this in which the meaning of a word is uncertain, translators often examine where and how each ancient author employed it. In the case of the Gospel of Philip, the Coptic word *koinwnoc* was used twice. In both instances, it is used to describe Mary's relationship with Jesus.[1048] In contrast, at every one of the four locations where the Gospel of Philip clearly refers to a marital partner, it uses the Coptic word *Chime*. Hence, if the author wanted to denote a marital relationship between Jesus and Mary Magdalene, he or she would have in all probability used this word.[1049]

In sum, the evidence is very much against the claim that "companion" meant "spouse."

What about Jesus kissing Mary on the mouth? In reality, the lone manuscript we have of the Gospel of Philip does not say this. Like many ancient manuscripts, it is damaged and some letters and words are missing. (See pictorial example on next page.) In such instances scholars attempt to reconstruct the text, but the results can be tentative. This is just such a case. Here is what the manuscript actually says:

> [. . . loved] her more than [all] the disciples [and used to] kiss her [often] on her [. . .].

The brackets indicate where some or all of the text is missing. The words in the brackets are those that the translator feels confident in reconstructing. The blank spots are uncertain. Where did Jesus kiss Mary? The scholar who translated this explicitly states it may be on her mouth, or feet, or cheek, or forehead. He also notes that the choice of the word "kiss" is unsure because "the Coptic construction found here is not normally used in this sense."[1050]

What does all the above tell us? Above and beyond the fact that there is no rational reason to accept the Gospel of Philip at face value, it doesn't even state what Dan Brown claims it does. In short, it is abject nonsense to claim that "the marriage of Jesus and Mary Magdalene is part of the historical record."

PROPAGANDA VERSUS REALITY

On NBC's *Today Show*, host Matt Lauer asked Dan Brown, "How much of this is based on reality in terms of things that actually occurred? I know you did a lot of research for the book."

Brown responded:

> Absolutely all of it. Obviously, there are—Robert Langdon is fictional, but all of the art, architecture, secret rituals, secret societies, all of that is historical fact.[1052]

Given that Lauer lobbed this softball question and didn't faintly challenge Brown's answer, a wide swath of the *Today Show* audience probably believed it. Likewise, the same reviewer who claimed *The Da Vinci*

Code contains "several doctorates' worth of fascinating history and learned speculation," also called it "brain candy of the highest quality."[1053] What does it say of a society and journalists who swallow such blatant and easily disproved propaganda?

Copyright © Istituto Papirologico "G. Vitelli"—Firenze

Example of a manuscript missing letters and words[1051]
(not the Gospel of Philip)

Let's step outside of the fantasy worlds often crafted by popular media and sum up the realities here. First, comprehensive evidence indicates the

Biblical texts have been passed down to us with remarkable accuracy, particularly those comprising the New Testament. Second, the New Testament Gospels were all written within the conceivable lifespans of people who personally knew Jesus.[1054] Third, within 30 years of Christ's time on earth, Christians were being tortured and put to death by the Roman Empire.[1055] [1056] Fourth, the New Testament Gospels were accepted as authoritative well before the time of the Roman Emperor Constantine. All of this evidence is consistent with the traditional Christian view that the Biblical Gospels are accurate representations of Christ's time on earth.

CHAPTER 6

HOSTILE WITNESSES, COSMOLOGY, AND BIOGENESIS

One of the main sources of skepticism about the Bible is its assertion that God created the universe, stars, earth, plant life, and animals. The prevalent view among scientists is that God didn't make any of this.[1057] Even in Christian circles, many believe God initiated the "big bang" and let natural processes take over from there. Hence, it has been said by some, including Georges Lemaître, the priest and physicist who formulated the big-bang theory,[1058] that the Bible's account of creation is symbolic and meant to teach spiritual truths—not to explain the origin of the universe.[1059]

Putting aside for a moment the issue of interpreting the Bible, if natural processes can provide a reasonable explanation for the universe and everything within it, one can easily appreciate why some people are atheists. Conversely, however, if natural processes cannot truly account for all that exists and all that happens, there is no choice but to acknowledge the existence of God or something that transcends nature. Moreover, if a supernatural power created the universe, this amounts to a miracle far more incredulous than any other in the Bible. A phenomenon of this scale makes events like the parting of the Red Sea and the healing of a blind person look trivial by comparison.

I have studied thousands of books, articles, and scientific papers that contain information relating to the points above. This chapter and the next three encapsulate that research. In addition to their factual density, what makes them unique is that virtually every argument is substantiated with evidence from a hostile witness or witnesses. To put it another way, these chapters cite evolutionists to make a case for creation.

Throughout this book, reputable and scholarly sources are commonly cited, many of which display no indication of partiality toward the Bible. As I continue to utilize highly credible sources, the practice of relying upon unsympathetic/hostile sources will now become extremely stringent. There will be, however, a few instances in which I cite scholarly opponents of evolution to challenge this theory, but in every such case, I make it plainly evident I am citing a critic of evolution. Also, for easy reference, the main text includes the publication date of every key quote.

Some evolutionists attempt to portray all who oppose their theory as biased or misinformed.[1060] There can be little doubt this is true in some cases, but I intend to demonstrate the polar opposite is also true—that the theory of evolution is based upon misinformation and prejudice. Let me be very clear that I am not questioning the sincerity or intellect of evolutionary scientists. Many of them are decent and brilliant people. I am, however, questioning their embrace of an assumption that gives them no other choice but to believe in evolution. And what is this assumption? Below are several quotes from some forthright evolutionists who explicitly and implicitly reveal it.

Notions Versus Truth

Robert Jastrow was the director of Mount Wilson Observatory and the first chairman of NASA's Lunar Exploration Committee. He established NASA's Goddard Institute for Space Studies (which he directed for 20 years) and "hosted more than 100 CBS-TV network programs on space science."[1061] [1062] In a book published in 1978, Dr. Jastrow made the following statement that cuts to the heart of this matter:

> There is a kind of religion in science; it is the religion of a person who believes there is order and harmony in the Universe, and every event can be explained in a rational way as the product of some previous event, every effect must have its cause; there is no First Cause.[1063]

In other words, the religion of science is that there is no God. When a scientist begins with this premise, an entire realm of inquiry is barred. What if in fact, reality sits behind these bars? As Kansas State University biologist Scott C. Todd writes in a 1999 letter to the journal *Nature*:

> Even if all the data point to an intelligent designer, such an hypothesis is excluded from science because it is not naturalistic.[1064]

Likewise, in a 1997 article, Richard C. Lewontin, Professor of Biology at Harvard University, writes:

> It is not that the methods and institutions of science somehow compel us to accept a material explanation of the phenomenal world, but, on the contrary, that we are forced by our *a priori* adherence to material causes to create an apparatus of investiga-

tion and a set of concepts that produce material explanations, no matter how counter-intuitive, no matter how mystifying to the uninitiated. Moreover, that materialism is absolute, for we cannot allow a Divine Foot in the door.[1065]

In a book published in 2001, molecular biologist Franklin Harold writes:[1066] [1067]

[W]e are compelled by our calling to insist at all times on strictly naturalistic explanations; life must, therefore, have emerged from chemistry.[1068]

George Wald, Nobel Prize recipient and former Professor of Biology at Harvard University, writes in a 1954 paper:[1069]

Most modern biologists, having reviewed with satisfaction the downfall of the spontaneous generation hypothesis, yet unwilling to accept the alternative belief in special creation, are left with nothing. I think a scientist has no choice but to approach the origin of life through a hypothesis of spontaneous generation.[1070]

The problem with such viewpoints is that they are founded upon assumptions, and even the smartest person in the world will arrive at a false conclusion if they begin with a false assumption. What needs to be done is to dig beneath these assumptions to see if they are warranted. Throughout the following chapters, we will examine evidence indicating that there is a First Cause, that the data does point to an intelligent designer, and that life did not spontaneously emerge from chemistry. And what of the view that science "cannot allow a Divine Foot in the door"? A far more logical view is that of George Washington Carver:

Science is simply the truth about anything. (1924)[1071]

Joseph Lister, inventor of the antiseptic procedures that ushered in the era of modern surgery,[1072] [1073] [1074] [1075] once said he found it "strange" how often the following basic principle was disregarded:

In investigating nature you will do well to bear ever in mind that in every question there is the truth, whatever our notions may be. (1876)[1076]

There is a growing tendency for people to equate their notions with truth. They speak in terms of "my truth" and "your truth." It's easy to embrace this perspective in our era of modern comforts, but how would this change for a person who found himself on an operating table in a nineteenth century surgical ward with a 45% mortality rate? No sane person would talk about "my truth" when it came to whether or not the scalpel was sterilized. Such thinking is a luxury of those who don't contend with hard realities because of their circumstances or unwillingness/incapacity to face facts. While recounting an event of his younger days in which he presented some findings to a teacher of "high reputation," Lister described the teacher's reaction with these incisive words:

> I was very much struck and grieved to find that, while all the facts lay equally clear before him, those only which squared with his previous theories seemed to affect his organs of vision. … When I was a little boy I used to imagine that prejudice was a thing peculiar to some individuals. But, alas! I have since learned that we are all under its influence. (1876)[1077]

How many times have you witnessed a debate that concludes with someone honestly stating, "I was mistaken. You are right and I am wrong"? Very few people have the fortitude to admit such a thing to themselves much less to someone else. For those who think that scientists are generally above this, consider what Max Planck, founder of the discipline of quantum physics stated:[1078]

> A new scientific truth does not become accepted by way of convincing and enlightening the opposition. Rather, the opposition dies out and the rising generation becomes well acquainted with the new truth from the start.[1079]

Likewise, Robert Jastrow writes:

> It turns out that the scientist behaves the way the rest of us do when our beliefs are in conflict with the evidence. We become irritated, we pretend the conflict does not exist, or we paper it over with meaningless phrases. (1978)[1080]

Herein lies a major reason evolution is accepted. As will be shown, it is a paradigm that was mainly established in the 1800s, and although its underpinnings have been destroyed by the advances of science, prejudicial notions prevent some people from questioning it. With this in mind, let us attend to some general but critical points regarding the evidences we will soon examine.

Quoting in Context

In our age of unprecedented access to information, it is shameful how often statements are taken out of context.[1081] Equally, it is disgraceful that certain people accuse others of taking quotes out of context when in fact they have not. This has become a common ploy of those caught with their foot in their mouth,[1082] so let's set the matter straight right now.

With very few exceptions, I have personally examined the context of every quote and citation in this book. This means I traced them back to the original sources and studied the surrounding verbiage and general venue. The only notable exceptions are several instances in which the original source was written in a foreign language. In such cases, I relied upon scholarly works with the presumption that the translators did their work honestly and precisely. Although I have been very careful in my work, I stake no claim to perfection, and if I have misinterpreted something, I will publicly correct it at www.rationalconclusions.com. There, I will also respond to the inevitable attacks that will be leveled at this work.

To make sure the following points are thoroughly clear, allow me to state a few things that should be obvious. When citing evolutionists, I am not implying they accept the same conclusions I draw from the information they present. If this were the case, they would be creationists instead of evolutionists. I cite evolutionists precisely because they oppose creationism, thereby eliminating any possibility they are biased toward this view. As we will see, this can be done because of critical disconnects between facts they impart and opinions they hold. Likewise, when I quote someone, this doesn't mean I accept everything they have ever written or said. I am citing a specific creditable point they have made—not pronouncing them to be an infallible authority. Those familiar with scientific publications will recognize this as the common standard.

Reliable Sources

It never ceases to amaze me how some people will blindly accept unsubstantiated claims if they align with their personal views but indignantly demand absurd standards of proof for anything that runs counter to them. Louis Pasteur displayed an extraordinary insight into human nature when he stated that "the greatest aberration of the mind is to believe a thing to be, because we desire it."[1083] A tell-tale sign of this aberration is the habit of dismissing sources for arbitrary reasons. When reputable sources make assertions that are unpalatable to some people, their way of dealing with this is not to research the matter with an open mind but to escape from it with any excuse they can muster.

One such excuse is to claim a source is not reliable simply because it is ten, twenty, or however many years old it may happen to be. No intellectually honest person labors under the notion that a source is reliable or unreliable merely because of its age. Papers in peer-reviewed academic journals, which are said to be "the gold standard of scientific credibility,"[1084] often cite sources that are more than ten, twenty, thirty, or even fifty years old.[1085] [1086] [1087] [1088] [1089] [1090]

Moreover, at least one prominent evolutionist has written that he and "many other biologists" learned to "think carefully about candor in argument … in case one was furnishing creationist campaigners with ammunition in the form of 'quotable quotes', often taken out of context." (1999)[1091] (Note that the accusation about being taken out of context was leveled without any evidence to substantiate it.) If "many" evolutionists have decided not to be candid, we can hardly expect them to republish all the information that weighs against their theory every seven years or whatever arbitrary timeframe someone proclaims to be "up to date." If a certain assertion is demonstrably obsolete, that is one thing, but it is not obsolete unless facts clearly prove it to be so. Simply declaring "it's old" rings hollow.

Credentials and the Scientific Method

With regard to the predictable criticism that I don't have enough formal education to challenge people with doctoral degrees, it should be pointed out that despite the impression some try to convey,[1092] [1093] [1094] numerous scholars have expressed serious misgivings about the theory of evolution. These include over 700 Ph.D. scientists who have signed a declaration of "scientific dissent" from Darwinism.[1095] Many of the overarching principles and details of this chapter are derived from the writings of such people, and credit is due to them even though their works are rarely cited herein because I am relying upon hostile sources. Nevertheless, even those who are well-read in these topics will find much new and compelling information in the pages that follow.

Let me be clear that I make no claim to being smarter or more educated than anyone cited or contested in this book. What I do claim to have done is diligently researched facts and followed them where they led. For those who may think this is what every scientist does, examine what famed Harvard paleontologist Stephen J. Gould wrote about science in general and evolution in particular:[1096]

The inadequacy, for example, of a "hard sciences" model for crucial experiments in proof and disproof has never been more evident. The data of natural history [i.e., evolution] are so multifarious, complex, and indecisive that simple accumulation can almost never resolve an issue. Counter-cases can always be documented in large numbers, and no one can find and count enough unbiased cases to establish a decisive relative frequency. Theory must play a role in guiding observation, and theory will not fall on the basis of data accumulated in its own light. (1977)[1097]

That's the problem right there: "theory will not fall on the basis of data accumulated in its own light." Because scientists start with an assumption that can only lead to one result, one can easily see how intelligent, educated, and sincere people can believe in evolution—and why no amount of hard data or counter-examples will allow it to fall in their minds. Join me now as we step outside the self-imposed confines of prejudice under which many labor, and take a hard look at the theory of evolution starting with the big bang and continuing to all that exists today.

The Big Bang

A colleague once said to me that the big bang is a scientific fact. I replied, "Can you tell me what the scientific evidence for it is?" His answer was, "The evidence runs over my head, but science has proven it. All these scientists can't be wrong."

This type of mindset is typical and quite understandable when it comes to dealing with complex subjects like cosmology. Most of us don't have the time or mathematical ability to fully understand the intricacies of such matters. Instead, we rely on people who do such things for a living to explain them to us in general terms. What happens, however, when scientists disagree with each other? Do we simply count votes and go with the majority opinion?

Although textbooks, journals, newspapers, magazines, and documentaries often portray the big bang as the only well-founded theory for the formation of the universe, notable scientists have expressed strong skepticism regarding central aspects of it, and some have outright rejected it. For example:

Hannes Alfvén, winner of the 1970 Nobel Prize in Physics; Professor of Theoretical Electrodynamics, Electronics, and Plasma Physics at the Royal Institute in Sweden, Professor of Electrical

Engineering at the University of California, San Diego:

> [T]here are an increasing number of observational facts
> which are difficult to reconcile in the Big Bang hypothesis.
> The Big Bang establishment very seldom mentions these,
> and when nonbelievers try to draw attention to them, the
> powerful establishment refuses to discuss them in a fair way.
> … The Big Bang is indeed a cosmology of the same charac-
> ter as the Ptolemaic [sun, planets and stars revolving around
> the earth]: absolutely sterile. (1984)[1098]

Halton Arp, Ph.D. in Physics and Astronomy from the California
Institute of Technology, staff astronomer at the Mt. Palomar and
Mt. Wilson observatories for 28 years, senior research scientist
with the Max Planck Institute for Astrophysics:

> I believe that the big bang theory should be replaced, because
> it is no longer a valid theory. (2006)[1099]

Geoffrey Burbidge, Ph.D. in Theoretical Physics from the University of London, director of the Kitt Peak National Observatory,
professor of physics at the University of California, San Diego:

> [A]stronomical textbooks no longer treat cosmology as an
> open subject. Instead the authors take the attitude that the
> correct theory has been found. … The situation is particu-
> larly worrisome because there are good reasons to think that
> the big bang model is seriously flawed. (1992)[1100]

Ari Brynjolfsson, Ph.D. and post-doctorate in nuclear physics
from the Niels Bohr Institute, University of Copenhagen (Den-
mark):

> [The supernovae] data indicate that the contemporary big-
> bang hypothesis is false. (2006)[1101]

None of the people above are creationists.[1102] [1103] [1104] [1105] They oppose
the big bang theory simply because they do not think it is consistent with
scientific evidence. In addition to such individuals, a number of Christian
physicists have concluded that the big bang theory is flawed and that the

Bible contains a literal and accurate record of the universe's creation. These include:

D. Russell Humphreys, Ph.D. in Physics from Louisiana State University

Jason Lisle, Ph.D. in Astrophysics from the University of Colorado

Eugene F. Chaffin, Ph.D. in Theoretical Nuclear Physics from Oklahoma State University

Danny R. Faulkner, Ph.D. in Astronomy from Indiana University

Don DeYoung, Ph.D. in Physics from Iowa State University

Keith Wanser, Ph.D. in Condensed Matter Physics from the University of California, Irvine

John Baumgardner, Ph.D. in Geophysics and Space Physics from the University of California, Los Angeles

John Rankin, Ph.D. in Mathematical Physics from the University of Adelaide (Australia)

Charles W. Harrison, Ph.D. in Applied Physics from Harvard University[1106]

In May of 2004, 34 scientists (including three of the evolutionists mentioned above) published an open letter in the magazine *New Scientist* criticizing the big bang theory. Among other things, the letter offers the following assessment:

> The big bang today relies on a growing number of hypothetical entities, things that we have never observed…. Without them, there would be a fatal contradiction between the observations made by astronomers and the predictions of the big bang theory. In no other field of physics would this continual recourse to new hypothetical objects be accepted as a way of bridging the gap between theory and observation.[1107]

Keep these words in mind as we move forward. You will see exactly what these scientists are talking about. As we delve into specifics, it is important to realize that practically all of the sources cited below are people who are strongly supportive of the big bang theory. In fact, many of them played instrumental roles in advancing it.

THIS IS NOT YOUR FATHER'S BIG BANG

What does the big bang basically involve? According to a college astronomy textbook published in 1978, such theories:

> say that once upon a time there was a great big bang that began the universe. From that moment on, the universe expanded, and as the galaxies formed they shared in the expansion…. [G]ravity has been slowing down the expansion. … If gravity is strong enough, then the expansion will gradually stop, and a contraction will begin. If gravity is not strong enough, then the rate of expansion might slow, but the universe would continue to expand forever….[1108]

If this is how you view the big bang, be prepared for a few surprises because some major changes have occurred. Let's lay a little groundwork to help us understand why. If the universe began with an explosion that took place billions of years ago, some very unique conditions are necessary to produce a situation capable of sustaining life. This is because if the outward force of the big bang was not great enough, the force of gravity would quickly cause the universe to collapse. But if it was too strong, the contents of the universe would spread out so quickly that galaxies, stars and planets could not possibly form.[1109] We're speaking in generalities, but scientists have put hard numbers to this. As renowned physicist and big bang pioneer Robert Dicke explained in 1969,[1110] if the speed at which the universe expanded during the big bang differed by 1/10th of one percent, the universe would have either collapsed before stars could form, or would now be expanding 3,000 times faster than it currently is. In summary, Dicke remarked:

> There seems to be no fundamental theoretical reason for such a fine balance.[1111]

▶A universe balanced in this manner is said to be "flat" and hence, this is called the "flatness problem." To get a little more technical, consider the following: In trying to determine whether or not the universe will expand forever or recollapse, scientists measure the approximate mass density and expansion rate of the universe and then use these parameters to perform calculations. The result is that the universe is found to be balanced on a razor thin borderline between expanding forever and recollapsing. This is defined through a "density parameter" called omega or Ω. Omega is the density of the present universe divided by the density that represents the borderline between expanding forever and recollapsing, which is called the "critical density."

Ω = <u>Density of the universe</u>
Critical density

If $\Omega = 1$, this means the density of the universe is equal to the critical density and the universe is perfectly balanced between expanding forever and recollapsing. As of 1997, observations showed that Ω was somewhere between 0.1 and 2. This may seem to be an appreciable range, but when we extrapolate these measurements back billions of years to the supposed big bang, we find that for Ω to fall between 0.1 and 2 at present, it had to be almost precisely 1 at one second after the big bang. What does "almost precisely" mean? It means no further from 1 than 0.000000000000001.[1112] [1113] [1114] ◄

Here is where our first major change steps onto the scene. In December of 1979, a physicist by the name of Alan Guth was attempting to resolve another issue related to the big bang theory (called the magnetic monopole problem). In doing so, he found a potential solution that had the effect of causing the universe to multiply in size by roughly 10^{55} times in less than a trillionth of a second (10^{55} is shorthand for the number 1 with 55 zeros after it.) A natural consequence of this event would be to drive the universe to the state of "fine balance" we have been speaking of.[1115] [1116]

No need to get into the details, but this idea, called "inflation," was rapidly embraced by the scientific community. Why? In the words of Richard Morris (Ph.D. in theoretical physics from the University of Nevada):

> [Inflation] has gained such wide acceptance not because that it had been confirmed by experiment, but rather because no one has been able to think of any other reasonable way that a universe like ours could have evolved. (1993)[1117]

Similarly, when asked why inflation caught on so quickly, renowned cosmologist Jim Peebles said that:

> we didn't have any other options. ... I think we should be careful. I think there's a reasonable chance we've been led down the wrong path. It certainly has happened before. (1988)[1118]

Some would argue that new observational evidence has confirmed inflation,[1119] but there are good reasons to consider such claims exaggerated.[1120] [1121] Moreover, as the quotes above demonstrate, the theory was quickly embraced without any real evidence to support it. Because the concept of inflation plugged up holes in the big bang theory, scientists considered this evidence enough. We are going to see this pattern repeated again and again.

Even more dubious than scientists' eager acceptance of inflation is the fact that no known natural phenomenon exists that could cause such a thing to take place. Remember, we are talking about an expansion of the universe by a factor of 10^{55} in a split second. To get an idea of the immensity of this number, realize there are roughly 10^{80} atoms in the entire known universe.[1122] Surely, such a theory should at least have a plausible explanation behind it. Dr. Guth came to the idea of inflation through pursuing a hypothesis in a branch of science called particle physics, but it was later determined that his model was incompatible with real-world observations.[1123] [1124] [1125] [1126] Since that time, many other inflation scenarios have been proposed,[1127] but as Dr. Guth explains:

> We still appear to be a long way from pinning down the details of the particle physics that underlies inflation. (1997)[1128]

Likewise, the college textbook *Foundations of Modern Cosmology* states:

> Inflation also does not seem to fit comfortably within any known scenario of particle physics. ... Replacing one set of ad hoc requirements, special initial conditions, with another, a particle that seems to have nothing else to do with particle physics, at least not yet, does not seem all that satisfying. (1998)[1129]

Shortly before Dr. Guth arrived at the idea of inflation, a Russian scientist by the name of Andrei Linde came to the same idea and discarded it because he foresaw the issue that would later sink Dr. Guth's model, but perhaps more importantly, he did not realize that it had the potential to resolve some major problems with the big bang theory.[1130] After Dr. Linde became aware that inflation had this benefit, he went back to the drawing board and developed other models for it. In describing why he decided to pursue a theory he initially rejected, Dr. Linde said:

> [I]t was very difficult to abandon this simple explanation of many different cosmological problems. I just had the feeling that it was impossible for God not to use such a good possibility to simplify his work, the creation of the universe. (1987)[1131]

When asked about this statement and if he thought that the universe "was set up by some intelligence," Dr. Linde replied:

> As for myself, I cannot say that I am religious. But I am also not a straightforward materialist who believes that everything is mat-

ter and nothing else. It is also very dangerous to write about such things in the press, but in my opinion we have overlooked something very important. (1987)[1132]

This statement is stunning and quite revealing given that the concept of inflation rings of something that is repeatedly affirmed in the Bible, which states that God:

> created the heavens and stretched them out (Isaiah 42:5) … stretched out the heavens (Isaiah 45:12) … stretched forth the heavens (Isaiah 51:13) … stretched out the heavens by his discretion (Jeremiah 10:12) … stretched out the heaven by his understanding (Jeremiah 51:15) … stretcheth forth the heavens (Zechariah 12:1) … alone spreadeth out the heavens (Job 9:8)[1133]

In summary, the dominant scientific theory for the origin of the universe doesn't consist of just a big bang anymore but of a big bang that produced a very short period of regular expansion, followed by inflation that produced a very short period of super-rapid expansion, followed by a lengthy period of regular expansion.

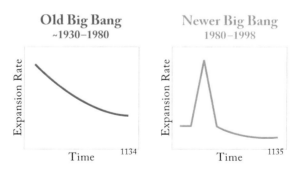

Anti-Gravity: The Cure for the Big Bang Blues

Let's move away from what allegedly happened billions of years ago and deal with what is taking place at present. The following statement appears in a 1977 book written by Steven Weinberg, a Harvard University professor who was awarded the Nobel Prize in physics two years later:[1136]

> The galaxies are not rushing apart because of some mysterious force that is pushing them apart…. Rather, the galaxies are moving apart because they were thrown apart by some sort of explo-

sion in the past.[1137]

Many other scientists have written basically the same thing,[1138] [1139] [1140] but I chose to highlight this quote because the author also makes his point in a negative sense by saying that a "mysterious force" is not causing the universe to expand. This is ironic because the exact opposite stance is now an integral part of the big bang theory. In fact, while being interviewed for a 2005 PBS documentary, Dr. Weinberg takes the existence of this force for granted and even uses the word "mysterious" to describe it.[1141] Various names have been given to this force including "lambda," the "cosmological constant," "vacuum energy," "dark energy," and "quintessence." We're going to avoid these names and take the liberty of calling it something that makes intuitive sense: anti-gravity.[1142]

Some history will illuminate the situation. In 1916, Einstein published his theory of general relativity, which informs us about the nature of gravity. A year later, he published a paper in which he used his equations of general relativity to develop a scientific model of the universe. At that time, observations of stars showed they didn't move much, and Einstein used this "fact" as one of the starting assumptions in his model. But when he began making calculations, Einstein ran into trouble. Despite the complexities of general relativity, this is pretty easy to understand. Gravity pulls things together. In the absence of any other force, the contents of the universe should be attracting each other and getting closer. Yet, observations indicated this was not happening. Given this conflict, Einstein surmised there must be some sort of unknown force that counter-balanced gravity and stabilized the universe. Though he admitted that such a force was "not justified by our actual knowledge of gravitation," he proceeded to add an expression for it into his previously published equations of general relativity.[1143] [1144] [1145] [1146]

By 1931, however, newer observations had convinced Einstein that the universe was expanding. Since Einstein invented this concept of anti-gravity to explain why the stars were fairly stationary, he discarded it.[1147] Not only that, but he later referred to it as "theoretically unsatisfactory" and told a colleague it was the biggest mistake he had ever made.[1148] [1149]

Although Einstein was finished with anti-gravity, others were not. In the early 1930s, scientists used measurements of the expansion rate of the universe and the distances between galaxies to calculate a date for the big bang. The answer was 1.7 billion years ago. The result was straightforward and clear, but it created a dilemma because radioisotope dating produced an age for the earth of about 3.5 billion years. How could a 3.5 billion-year-old planet exist in a 1.7 billion-year-old universe? One of the primary rationalizations placed on the table was the idea of anti-gravity. Since it would cause the expansion rate to be faster in the present than it was in the past, this would allow the universe to be older than the calculation showed. The big bang theory lived with this paradox for more than 20 years, but in the 1950s, scientists determined there was far more distance between the galaxies than previously thought. Because there was no longer any need for anti-gravity to explain away the age discrepancy, it was promptly dropped. As one of the great physicists of that era described it, this discovery "killed" the idea of antigravity.[1150] [1151] Notice the pattern yet?

Fast forward to 1994 when a new conflict arose. By this time, scientists had concluded that the oldest stars were at least 12 billion years old. Yet, modern measurements of the expansion rate of the universe and the distances between galaxies showed the universe could be no more than 11 billion years old. Moreover, by this time inflation had become generally accepted, and when it was taken into account, the calculated age of the universe was estimated to be 8.1 billion years. Thus, "science" was left with 12 billion year old stars in an 8 billion year old universe. As you might suspect, talk ensued about taking anti-gravity out of the closet again.[1152] Then in 1998, two teams of scientists announced even newer measurements indicating the rate of expansion was speeding up.[1153] [1154] Wait a minute. Wasn't the force of gravity supposed to be slowing down the expansion rate? Absolutely. Again, we quote from Dr. Weinberg in his 1977 book:

> [T]he galaxies have not been moving apart at constant velocities, but have been slowing down under the influence of their mutual gravitation.[1155] [1156]

So how could the galaxies be speeding up? Anti-gravity of course, and within several years this idea became generally accepted in the scientific community.[1157] [1158] It is claimed that the idea of anti-gravity shouldn't be considered "theoretically unsatisfactory" as Einstein said it was because there is now a physical explanation for it. Something called the "standard model of particle physics" predicts that anti-gravity should exist.[1159] [1160] [1161] [1162] [1163] However, this same model also predicts that its strength should be 10^{120} times stronger than observations allow for.[1164] [1165] This enormous level of inconsistency amounts to no explanation at all.

Observe the mindset. The big bang theory cannot be wrong, and therefore, any non-falsifiable idea that works in its favor must be true even if it lacks a physical explanation or has extreme inconsistencies. To summarize, when the latest observations are interpreted in the context of the big bang theory, there is a clash with the law of gravity. Yet, instead of questioning the underlying framework of the big bang model, a new "mysterious" force has been embraced: anti-gravity. A NASA website adequately sums up the situation:

> The main attraction of [anti-gravity] is that it significantly improves the agreement between theory and observation. (2008)[1166]

Consequently, this is how the big bang theory looks now:

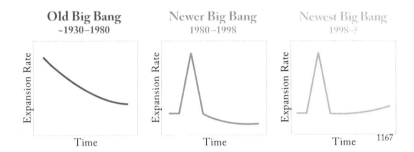

And serious problems are still lurking. The person who in 2000 was considered to be "the world's foremost theoretical cosmologist" is Jim Peebles of Princeton University. His research was instrumental in developing two of the three primary evidences cited in support of the big bang theory, and his accomplishments have been recognized with prestigious academic awards and honors.[1168] [1169] [1170] [1171] [1172] In September of 2002, Dr. Peebles gave a lecture he entitled "Problems with the Standard Cosmological Model."[1173] In common language, this means "Problems with the Latest Rendition of the Big Bang Theory." I extensively cite this lecture below, and though it may be tempting to skim such lengthy quotes, I highly recommend soaking in every word. Dr. Peebles begins gently and makes a joke about being advised against "raising doubts":

> So, first, I've got to explain that the title and abstract were constructed in a big hurry.... It came out a little stronger than I intended. I am very impressed with the successes of the now-standard cosmology. What I intend to describe are some, perhaps, details, aspects of the phenomenology that to me, just don't seem to make sense within the standard model. I think they're serious enough problems that they are going to drive adjustments of the standard cosmology, perhaps very simple fine tuning, perhaps

something more serious. We'll see. … Now one final point on discussion. I am told that I may not take as my first goal that of annoying all of my colleagues in cosmology by raising doubts, but rather it is to entertain the rest of you people. [Laughter.]

As the talk progresses, Peebles becomes very blunt. We're not going to delve into the technical details, but for those who are interested, a video of the lecture is currently available online.[1174] Instead, we will peruse an assortment of remarks that all pertain to discrepancies between the big bang theory and real-world observations:

- "What in the world is going on here?"
- "It's obscene."
- "[I]t's such a bizarre phenomenon. I can't believe it comes out of the standard cosmology."
- "I've not been able to get any sensible person to actually look because it's such a ridiculous phenomenon."
- "There are even more details that I maybe hesitate to get into in a public conversation such as this, but let me just mention one of them."
- "It's ugly. Let us leave the situation at this—a feeling of nervousness at least on my part.…"
- "[O]ne clear and present danger.…"
- "The late galaxy formation that has been so heavily advertised in the standard cosmology, I think didn't happen. It's not there. Give me a break."
- "That's not the way it is. Am I over the top? Isn't there a discrepancy here between theory and observation?"
- "Let's go on to another aspect of galaxy formation that again, seems to me, fishy."
- "I didn't get around to making a transparency because it's too depressing. The model doesn't match the real world.…"
- "We have then just another example of something that just doesn't make sense to me. The phenomenon seems so reproducible and so out of line with what I would expect from this as the realization of our world."
- "People don't worry about it because they don't think about it. They aren't nervous for reasons that totally escape me.…"
- "[I]t is weird. It is weird.…"

• "There's no pattern I can see. I'm frustrated."

Given Dr. Peebles' credentials and the provocative content of this lecture, one might think it would have garnered a huge amount of attention. Precisely the opposite. Searches performed in May 2006 through Google and Lexis-Nexis netted a total of one source that contains any record of it or reference to it. This comprises a single link to an audio file of the talk at the academic institution where it was given. Based on my research, there are no transcripts, commentaries, articles, blogs, or anything else about this presentation available. According to Google, there is no other web site in the world that even links to this presentation.

Furthermore, the issues brought out by Dr. Peebles are just the tip of the iceberg. Even if these problems are resolved without further disfigurement of the big bang theory, a host of others exists, some of which have been sidestepped for years and others that are emerging as more detailed astronomical observations are made. These problems include but are not limited to the entropy problem,[1175] the energy problem,[1176] [1177] the anti-matter problem,[1178] [1179] [1180] [1181] [1182] [1183] [1184] [1185] the young galaxy problem,[1186] [1187] [1188] the mature galaxy problem,[1189] [1190] the shadow problem,[1191] and most notoriously, the "What came before the big bang?" problem.[1192] [1193] [1194] In a book published in 2001, Dr. Weinberg writes that we can be "quite confident" about the big bang theory, yet four pages later he admits:

> As we make progress in understanding the expanding universe, the problem itself expands, so that the solution seems always to recede from us.[1195]

Such circumstances are the hallmark of an erroneous theory. Consider these words from the aforementioned textbook, *Foundations of Modern Cosmology*:

> Pseudoscience is often based on observations and may cite much "confirming evidence" but never permits refutation. Either the contrary data are ignored, or new details are continuously added to the theory in order to explain all new observations. (1998)[1196]

I leave it to you to determine whether or not this statement bears any resemblance to the big bang theory. Nevertheless, let's accept the theory for the time being. Let's put aside the fact that it has needed repeated modifications to keep it from collapsing in the face of new evidence and is still rife with problems and discrepancies. Let's grant evolution the benefit of a doubt and just believe it for now.

BIOGENESIS

Now we can proceed to the next stage of evolution, which is moving from nonliving matter to living. Dr. Francis Crick (1916–2004) was one of the two people responsible for discovering the structure of DNA, an accomplishment regarded as the "greatest biological advance of the twentieth century." Beyond this, he made other momentous contributions to the science of biology and was honored with many impressive titles and awards.[1197] [1198] [1199] Crick was a staunch and vocal atheist. He specifically stated one of the main reasons he went into science was to tear down religious beliefs by bridging the gap between living and nonliving matter.[1200]

In 1981, Crick published a book entitled *Life Itself*, which was formed around the premise that aliens may have sent a spaceship filled with microorganisms to earth to get life started here.[1201] In this book, he speculates about their circumstances and motivations, devotes ten pages to the design of their spacecraft, and states that they might be secretly watching us.[1202] What would prompt someone with such remarkable scientific credentials to undertake such wild speculation? He wasn't a UFO junkie and dismissed alleged sightings as "probably without significance." Although he treats the spaceship theory very seriously in his book,[1203] Crick later told a colleague his main purpose was to get "intelligent" people thinking about "the *problem*." What *problem*? This is best explained by his own words in the book:

> An honest man, armed with all the knowledge available to us now, could only state that in some sense, the origin of life at the moment appears to be almost a miracle, so many are the conditions would have had to been satisfied to get it going.

No atheist would let such an admission stand alone, and he immediately follows this up by stating:

> But this should not be taken to imply that there are good reasons to believe that it could *not* have started on earth by a perfectly reasonable sequence of fairly ordinary chemical reactions. The plain fact is that the time available was too long, the many microenvironments on the earth too diverse, the various chemical possibilities too numerous and our knowledge and imagination too feeble to allow us to be able to unravel exactly how it might or

might not have happened such a long time ago, especially as we have no experimental evidence from that era to check our ideas against.[1204]

The tension between these two adjoining thoughts is obvious and is further reinforced by other statements he makes in the same book, such as the declaration that when we strip life down to its most basic elements:

> we cannot help being struck by the very high degree of *organized complexity* we find at every level....[1205]

> Every time I write a paper on the origins of life I swear I will never write another one, because there is too much speculation running after too few facts....[1206]

Actually, there are plenty of facts we will examine in the upcoming pages. The problem for Crick is that they are at odds with the notion that life could have arisen by natural processes.

Spontaneous Generation

In our modern era, a basic tenet of science is that life does not spring from non-living matter, but this was not always the case. As a high school biology textbook explains:

> From the time of the ancient Greeks until well into the nine-teenth century, it was common "knowledge" that life arose from nonliving matter all the time. Many people believed, for instance, that flies came from rotting meat, fish from ocean mud, frogs and mice from wet soil, and microorganisms from broth. Experiments performed in the 1600s showed that relatively large organisms, such as insects, cannot arise spontaneously from nonliving mat-ter. However, debate about how microscopic organisms arise con-tinued until the 1860s. In 1862, the great French scientist Louis Pasteur confirmed what many others had suspected: All life today, including microbes, arises only by the reproduction of preexisting life. (1997)[1207]

This is a typical description of this episode in history, but it is not entirely accurate. The dispute about spontaneous generation continued well beyond Pasteur's famous experiments of the early 1860s. One of pri-mary reasons for this was the publication in 1859 of Charles Darwin's landmark book about evolution, the *Origin of Species*.[1208] In his first edition,

Darwin writes life was "breathed" into the first organisms,[1209] and in the second edition, he goes further by affirming life was "breathed" into the first organisms "by the Creator."[1210][1211] However, a personal letter written by Darwin a few years later reveals that he didn't really believe this:

> But I have long regretted that I truckled to public opinion, and used the [Biblical] term of creation, by which I really meant 'appeared' by some wholly unknown process. (1863)[1212]

Likewise, many of Darwin's leading supporters felt the same way:

- Ernst Haeckel, a prominent biologist who was "one of Darwin's most ardent supporters":[1213][1214]

  > When Darwin assumes a special creative act for his first species, he is not consistent at any rate and, I think, it is not intended to be taken seriously. (1862)[1215]

- Edmund Perrier, professor of zoology and director of the Museum of Natural History in Paris:[1216]

  > [S]pontaneous generation is the foundation of the doctrine of evolution. (1879)[1217]

- Karl von Nägeli, a leading evolutionist, botanist, and cell biologist:[1218][1219][1220]

  > To deny spontaneous generation is to proclaim a miracle. (1884)[1221]

Another related factor that prolonged the spontaneous generation dispute was a mistaken notion about the simplest microscopic life forms. In the *Origin of Species*, Darwin writes that such "animals" are "composed of a gelatinous material, and show scarcely any trace of distinct organs."[1222] It is reasonable to believe that something so simple might spontaneously generate, but Darwin was oblivious to how complex and remarkable these creatures are.[1223] Despite the fact that they are far less intricate than higher life forms, there is currently no description, diagram, or picture that can come close to illustrating the full extent of their complexity. Take the example of an *E. coli* bacterium, which is the "simplest organism about which we know the most." Scientists are attempting to construct a computer model of one of these creatures. The task requires so much computing power that researchers are only planning to model about one quarter of it, and even so, they estimate it will be another five to ten years before "computers will be able to deal with this." (2005)[1224] This is just for a computer model, much less a real organism.

In Pasteur's and Darwin's era, the issue of spontaneous generation was not just an academic debate. It had life or death implications. Transmissible and highly fatal diseases like childbed fever ravaged Europe, and doctors who adhered to the misguided view that such diseases could spontaneously generate were often the very people responsible for transmitting them. Pasteur issued warnings that this was the case, but they went mostly unheeded.[1225] [1226] The tremendous frustration this caused him is evident in a letter he wrote to a medical scholar who was a leading supporter of spontaneous generation:

> Do you know why I desire so much to fight and conquer you? It is because you are one of the principle [adherents] of a medical doctrine which I believe to be fatal to progress in the art of healing—the doctrine of the spontaneity of all diseases. (1877)[1227] [1228]

Thankfully, today we've been cured of this ancient superstition, or have we?

Life in a Test Tube?

If life does not spontaneously generate from non-living matter, where did it come from in the first place? According to numerous sources that address this topic, the conditions that prevailed on earth in the distant past enabled life to form.[1229] What evidence do these sources cite for this claim? Almost universally, they cite experiments in which organic molecules are made from non-organic ones using "possible primitive earth conditions." Most famous among these is the Miller-Urey experiment of 1953, in which the "building blocks of life" (amino acid molecules) were made from inorganic substances.[1230] It is claimed that such organic materials accumulated over time into a "soup" or "broth" from which life arose. What is typically missing from textbooks, however, is an explanation of the pitfalls associated with these experiments.[1231] These include but are not limited to the fact that living organisms are constructed of "optically active" molecules, which are not the type of molecules produced in these experiments.[1232] [1233] [1234] Furthermore, the alleged "primitive earth conditions" are highly dubious, and the experiments involve an implausible degree of human manipulation.[1235] [1236] [1237] [1238]

Nevertheless, let's be acquiescent students, follow along in the textbooks and blindly disregard these problems as many scientists do. Let's take for granted that all of the right molecules necessary for life could have arisen naturally and found their way to each other. Reading on, we might get the impression that once the right organic substances are present, the emergence of life will surely follow. One textbook even refers to the Miller-Urey experiment as "life-in-a-test-tube." (1998)[1239] This is misleading in the extreme. Why? Because for more than 150 years, scientists have known there is no great divide between inorganic and organic molecules. In fact, chemists have been making one from the other since 1828.[1240] [1241] The primary barrier to life is not this but rather the amazingly complex manner in which these molecules are organized. To quote Franklin Harold, a molecular biologist who holds a Ph.D. in biochemistry from the University of California and has "40 years of experience in research on cell biology and microbiology":[1242] [1243]

> What distinguishes the cell from the soup is the former's purposeful organization; how strange then, to find the literature all but silent on the genesis of that organization! (2001)[1244]

There is an enormous difference between the molecules that make up a cell and the cell itself. This has been known since the days of Pasteur, who conducted experiments on liquids containing the remains of once-living organisms.[1245] In other words, all of the necessary raw ingredients of life were present at the outset. In fact, he went even further than this by using substances like blood that contained actual living cells, yet no new life emerged.[1246] [1247] Again, we quote from Dr. Harold:

> Biology textbooks often include a chapter on how life may have arisen from non-life, and while responsible authors do not fail to underscore the difficulties and uncertainties, readers still come away with the impression that the answer is almost within our grasp. My own reading is considerably more reserved. I suspect that the upbeat tone owes less to the advance of science than to the resurgence of primitive religiosity all around the globe, and particularly in the West. Scientists feel vulnerable to the onslaught of believers' certitudes, and so we proclaim our own. In reality we may not be much closer to understanding genesis than [scientists] were in the 1930s; and in the long run, science would be better off if we said so. (2001)[1248]

Additionally, in the same book he explains:

> Cell components as we know them are so thoroughly integrated

that one can scarcely imagine how any one function could have arisen in the absence of the others.[1249]

The origin of life appears to me as incomprehensible as ever, matter for wonder but not for explication.[1250]

Judging from these quotes, you might think Dr. Harold is a creationist, but the exact opposite is true. He is just very forthright in acknowledging that microbiology, his area of expertise, does not support the notion that life arose from nonlife.[1251] There are many other excellent points made by creationists in technical papers that pertain to this subject.[1252] [1253] [1254] [1255] While these are important matters for consideration, their complexity goes well beyond the scope of this book. Besides, there is one more realm of evidence we can look into that is far more easily understood. Moreover, it is the most compelling type of scientific evidence there is.

THE SOLE JUDGE OF SCIENTIFIC TRUTH

Richard Feynman (1918–1988) is described by the Encyclopædia Britannica as the "theoretical physicist who is widely regarded as the most brilliant, influential, and iconoclastic figure in his field in the post-World War II era." As people in the general public sometimes jokingly remark, "He's no Einstein," a popular saying amongst physicists was "He's no Feynman." Although he was an extremely creative thinker,[1256] Dr. Feynman was also a stalwart advocate of the view that science must be grounded in practical reality. He expressed this in the following manner:

The principle of science, the definition, almost, is the following: *The test of all knowledge is experiment.* Experiment is the *sole judge* of scientific 'truth'. (1963)[1257]

Once more, we go back to Louis Pasteur because this is the path he rigorously followed.[1258] In 1864, five years after the *Origin of Species* was published, Pasteur began a lecture at the University of Paris with these words:

Great problems are now being handled, keeping every thinking man in suspense; the unity or multiplicity of human races; the creation of man 1,000 year or 1,000 centuries ago, the fixity of species, or the slow and progressive transformation of one species into another; the eternity of matter; the idea of a God unnecessary. Such are some of the questions that humanity discusses nowadays.[1259]

While displaying his experimental apparatuses and a broth he placed into them, Pasteur described his experiments and concluded:

> And, therefore, gentlemen, I could point to that liquid and say to you, I have taken my drop of water from the immensity of creation, and I have taken it full of the elements appropriated to the development of inferior beings. And I wait, I watch. I question it, begging it to recompense for me the beautiful spectacle of the first creation. But it is dumb; dumb since these experiments were begun several years ago; it is dumb because I have kept it from the only thing man cannot produce, from the germs that float in the air, from Life…. Never will the doctrine of spontaneous generation recover from the mortal blow of this simple experiment.[1260]

More than 140 years have passed and untold numbers of pertinent experiments have been performed since this statement was made, and every word of it is as true today as it was then. In fact, some of Pasteur's flasks with their original broths still in them are on display in a Paris museum —and there they remain—lifeless to this day.[1261] Proponents of the notion that life emerged by natural causes can propose endless theories, but until they produce an experiment that creates life under natural conditions, the "sole judge of scientific truth" is decidedly against their belief.

One of the more recent theories to collapse in the face of experiment was the idea that hot spring waters containing clay provided a suitable environment for life to form.[1262] [1263] To quote the scientist who carried out the experiment:

> The results are surprising and in some ways disappointing. It seems that hot acidic waters containing clay do not provide the right conditions for chemicals to assemble themselves into 'pioneer organisms'. (2006)[1264]

Various reasons are offered to explain why life could arise in the past but not in the present. Typical among them is this excerpt from a biology textbook, which claims that it takes millions of years under certain conditions to generate life:

> On primitive earth, there were no bacteria to break down organic compounds. Nor was there any oxygen to react with organic compounds. As a result, organic compounds could accumulate over millions of years, forming that original organic soup. Today, however, such compounds cannot remain intact in the natural world for

a long enough period of time to give life another start. (2000)[1265]

Laboratories can easily simulate all of these physical conditions including a bacteria-free environment, so this is really only an issue of time. In a lab environment, we don't need millions of years for organic compounds to accumulate. We can insert them right from the start. That only leaves us with the time it takes for these compounds to react with each other. Besides the fact that the techniques of modern chemistry allow us to enormously increase the speed at which this occurs,[1266] I haven't seen scientists positing reactions that take a significant length of time. If they said, "Well, we have to wait for this reaction to occur and it takes too long to possibly test," that would be reasonable. But instead, they just throw out "time" as a general excuse. Scientifically, this doesn't cut it.

Another variant of the time excuse is the claim that anything can happen given enough time. One origin-of-life researcher put it this way:

> The most complex machine man has devised—say an electronic brain—is child's play compared with the simplest of living organisms. ... One only has to contemplate the magnitude of this task to concede that the spontaneous generation of a living organism is impossible. Yet, here we are—as a result, I believe of spontaneous generation. ... Time is in fact the hero of the plot. ... Given so much time, the "impossible" becomes possible, the possible probable, and the probable virtually certain. One only has to wait: time itself performs the miracles. (1954)[1267]

Absurd. This scientist readily admits that even the simplest living organism is far more complicated than a computer. NASA has had two robots crawling around on Mars for the past six years.[1268] Imagine if they came across a computer or even something as simple as a penny. It would be universally declared that it must have been made by an intelligent being. Anyone who claimed it was just a random "miracle of time" would be laughed at. Francis Crick aptly rejected such ridiculous notions:

> [I]t is not scientific to wave one's hands about and proclaim that in the long run all things are possible.[1269]

If any vaguely legitimate excuse exists for the failure of experiments to produce life, it is that there are countless possible ways in which molecules can interact with each other. This same reasoning, however, can just as well be used to argue that spontaneous generation is taking place right now, a notion that is almost universally rejected. Pasteur explained that one cannot rigorously prove a negative in cases like this,[1270] but there comes a point where reasonable people concede the obvious. To requote one of the biology textbooks above:

> All life today, including microbes, arises only by the reproduction of preexisting life. (1997)[1271]

Or Francis Crick:

> Pasteur … showed beyond doubt that in an initially sterile system, no sign of life would appear in even the richest and most tempting brew…. (1981)[1272]

Or another biology text book cited above:

> Louis Pasteur put an end to the spontaneous generation hypothesis. (2000)[1273]

Notice they don't say: "If Pasteur tried using different substances or waited a little longer, life would have arisen." Yet, his experimental findings are no stronger than the mass of experiments conducted since then that have failed to produce life under simulated primitive earth conditions. Any excuse that could be applied to these experiments can just as well be applied to Pasteur's. The clear implication of Pasteur's experiments is embraced by contemporary scientists. Why not also embrace the implication of modern experiments conducted under greater variety using superior technology? The unmistakable answer is that this would leave scientists no choice but to accept a supernatural explanation for the origin of life. Observe:

- George Wald, Professor of Biology at Harvard University and recipient of the 1967 Nobel Prize in Medicine:[1274]

  > Most modern biologists, having reviewed with satisfaction the downfall of the spontaneous generation hypothesis, yet unwilling to accept the alternative belief in special creation, are left with nothing. I think a scientist has no choice but to approach the origin of life through a hypothesis of spontaneous generation. (1953)[1275]

- Robert Shapiro, Ph.D. in organic chemistry from Harvard, postdoctoral training in DNA chemistry at Cambridge, and scientist with the Department of Chemistry at New York University:[1276]

  > Some future day may yet arrive when all reasonable chem-

ical experiments run to discover a probable origin for life have failed unequivocally. Further, new geological evidence may yet indicate a sudden appearance of life on the earth. Finally, we may have explored the universe and found no trace of life, or processes leading to life, elsewhere. In such a case, some scientists might choose to turn to religion for an answer. Others, however, myself included, would attempt to sort out the surviving less probable scientific explanations in the hope of selecting one that was still more likely than the remainder. (1986)[1277]

To discount the possibility that a divine being created life is a philosophy, not science. Given the current state of scientific knowledge (both experimental and theoretical), it is far less supportable to believe that life formed under natural conditions in the past than it was for certain obstinate scientists of the late 1800s to believe that spontaneous generation takes place in the present.

ARTIFICIAL LIFE

Incredibly, the impediments to evolutionary theory run even deeper. Beyond the inability to create life by simulating natural conditions, scientists have been unable to produce it artificially. For half a century now, it has been claimed that we are five to ten years away from creating life. Yet, here we are in 2009, and scientists have yet to produce even the simplest living organism using any and all means available to them.[1278] [1279] To quote one of the leading researchers in this field:

Here we are trying to understand the human genome with 24,000 some odd genes and 100 trillion cells and we don't know how 300 or 400 genes work together to yield a simple living cell. (2005)[1280]

This is in spite of the usage of sophisticated techniques such as scavenging parts from once-living organisms and injecting synthetic DNA into them. This high degree of intelligent manipulation is a far cry from the prospect of life emerging from a mindless soup of molecules. In the United States alone, 100 labs are working in this area, and one appears to be very close to success.[1281] [1282] [1283] I, for one, hope they succeed because the benefits could be enormous.[1284] Be assured that if this breakthrough comes, it will be hailed in some quarters as proof of evolution, but also realize that until life can be produced under conditions that reflect plausible natural circumstances, the notion that life arose spontaneously is an unscientific fable.

In the words of creation scientist Timothy G. Standish, Ph.D. in biology from George Mason University:

> Progressing in my studies, I slowly realized that evolution survives as a paradigm only as long as the evidence is picked and chosen and the great pool of data that is accumulating on life is ignored. … Only a small subset of evidence, chosen carefully, may be used to construct a story of life evolving from non-living precursors. Science does not work on the basis of picking and choosing data to suit a treasured theory. I chose the path of science which also happens to be the path of faith in the Creator. (2002)[1285]

Nevertheless, let's just blindly accept that life arose from non-living matter. Let's ignore the results of 140 years of experiments in which this has never happened and disregard the fact that scientists have thus far been unable produce even the most simple of life-forms using the latest revolutionary technologies. Let's believe that a cell, an organism far more complicated than a computer, spontaneously formed of its own accord. For the sake of argument, let's just swallow the discredited doctrine of spontaneous generation.

CHAPTER 7

GENETICS

The Science of Genetics Versus Darwin's Fables

Next, we attend to the topic of genetics, a field that involves the "scientific study of heredity and variation."[1286] An understanding of this extremely interesting discipline is critical if we are to evaluate the claim that evolution transformed simpler life forms into more complex ones. Darwin's theory asserts that over many generations, tiny singled-celled creatures evolved through many intermediate steps into human beings. The question thus arises: How did microbes acquire the genetic material to become people? When Darwin wrote the *Origin of Species*, the mechanisms of heredity were largely a mystery, and he states as much in the first chapter of this book:

> The laws governing inheritance are quite unknown; no one can say why the same peculiarity in different individuals of the same species, and in individuals of different species, is sometimes inherited and sometimes not so; why the child often reverts in certain characters to its grandfather or grandmother or other much more remote ancestor.... (1859)[1287]

In spite of this admission, Darwin voices some profoundly mistaken notions about this subject in the very same book, going so far as to call them "laws of inheritance."[1288] Although geneticists have known for decades that these ideas have no basis in reality, some of them persist among the general public to the present day. For instance, Darwin writes:

> I think there can be little doubt that use in our domestic animals strengthens and enlarges certain parts, and disuse diminishes them; and that such modifications are inherited. (1859)[1289]

In various places throughout the *Origin of Species*, Darwin offers specific illustrations of this idea. He believed if you trained an animal to perform a task, the effects of this training would be inherited by its offspring. For instance, if you consistently milked a cow, this would cause its descendants to have larger udders. When he observed crabs without eyes living in dark surroundings, he reasoned that their eyes had withered away over many generations due to lack of use.[1290]

The science of genetics has proven all of these contentions to be false.

The DNA of reproductive cells is not shaped by other body cells, and experimentation has clearly demonstrated that use and disuse have no effect on heritable traits.[1291] [1292] [1293] [1294] [1295] If you and your spouse exercise every day for ten years before you conceive, your child will not genetically inherit the results of your exercise. The same applies to any other environmental factor or condition. As articulated by a prominent developmental biologist in a book published by Oxford University Press:

> The powerful muscles of the blacksmith's arms are not inherited by his children; a mother's knowledge of Russian is not inherited by her children; giraffes did not acquire long necks by their ancestors stretching their necks to the highest branches. Characteristics and attributes acquired by experience or learning are not passed on to the offspring. The reason is simple. There is no mechanism whereby the acquired character—strong arms, Russian—can be transferred to the [reproductive] cells and appropriately alter their genetic constitution. (1991)[1296] [1297]

Note that recent study in a field known as epigenetics has found that environmental factors can alter the expression of genes in ways that are heritable,[1298] [1299] but in the words of a Professor of Biology at Harvard who specializes in this field, this merely "expands the range of options available to genes…. [T]he effects of use and disuse are not inherited…." (2007)[1300]

Darwin's view that use and disuse cause heritable changes can be traced to the French biologist Jean-Baptiste de Lamarck. More than 50 years before Darwin published the *Origin of Species*, Lamarck put forward the first real theory of evolution.[1301] [1302] [1303] It is amusing to observe the manner in which some modern writers try to distance Darwin from Lamarck.[1304] [1305] [1306] [1307] I can only surmise this is because Lamarck is very much associated with the discredited concept of use/disuse, while Darwin is the icon of evolution. Although Darwin did not agree with all of Lamarck's ideas, he followed in the same basic footsteps on two major areas: asserting that heritable effects were caused by use/disuse and claiming that all living creatures evolved from common ancestors.[1308]

Although the myth of use and disuse was debunked in the late 1800s,[1309] some scientists continued to espouse it well into the 20th century. One of these was the Marxist biologist Trofim Lysenko, who "dominated Soviet genetics and agriculture" for 24 years under Stalin and Khrushchev. Like Darwin, Lysenko thought that acquired traits could be inherited. Unlike

Darwin, however, Lysenko had the political authority to put this theory into practice on a nationwide scale. What was the result? In the words of the college textbook, *Principles of Genetics,* by the end of Lysenko's tenure in 1964, Soviet agriculture "was in shambles" and their "genetic research was an international disgrace."[1310] (1997)

Natural Selection Is a "Process of Elimination"

In Darwin's final and definitive edition of the *Origin of Species* (1872), he explains how an animal such as a horse "might be converted into a giraffe," proposing two mechanisms whereby this could happen. One was the "inherited effects of the increased use of parts." In other words, animals that stretched their necks to reach for food would pass along the effects of this stretching to their children. As we have already seen, this is an empty superstition. The other was natural selection. Darwin reasoned that animals that happened to have necks that were "an inch or two" longer than others would be able to reach food in high places that shorter animals could not. Thus, during famines, taller animals were more apt to survive and pass along the trait for longer necks to their children. This process, repeated over the course of many generations, would cause minor differences in neck length to accumulate and produce an increase of great magnitude.[1311]

Darwin argued for this idea based upon analogy with domesticated animals and plants. Domesticated creatures are those that have been cultured by humans for various reasons. Take, for instance, horses that are bred for speed by mating race champions. This is called artificial selection, as humans are selecting those traits deemed desirable for whatever purpose they have in mind. This is done with cats, pigs, cows, corn, broccoli, oranges, etc. Because the science of genetics was still unknown, Darwin relied heavily upon observations of such life forms.[1312] [1313] He reasoned that if artificial selection could affect populations in short periods of time, there was "no limit" to the amount of change that could be caused by natural selection over long ages.[1314]

Textbooks sometimes reinforce this notion by uncritically relaying Darwin's claim,[1315] but they fail to point out that this idea was discredited by the science of genetics almost a hundred years ago.[1316] If bacteria are to transform into humans, new genetic information must be created. Breeding domesticated animals does not do this and neither does natural or artificial selection. In fact, the changes that result from these processes are often caused by the loss of genetic information. Consider the manner in which humans have used artificial selection to create numerous breeds of dogs over the past century.[1317] For instance, if you would like to have big dogs, you select the largest ones and breed them. The same goes for breeding dogs with pug noses, certain colors, friendliness, sheepherding ability, and other traits.

http://en.wikipedia.org/wiki/Image:IMG013biglittledogFX_wb.jpg

Just from eyeing various breeds, one can easily get the false impression that artificial selection creates new genetic materials, but what actually takes place is the mixing and loss of existing genetic materials. The mixing aspect is very easy to understand. If two different breeds of dogs are mated,[1318] [1319] their offspring will inherit genetic material from both breeds. This does not produce new genetic elements and has no capability of turning a dog into an elephant or any other type of creature. What is not so obvious is that domestic breeding, especially when one is selecting for a particular trait such as size, often eliminates genetic information.[1320] As explained in a book written by two professors of genetics at Washington State University:

[I]n our efforts to produce animals and plants with just the desired characteristics, our domesticated animals and crops now have limited genetic diversity. This is due to the fact that, by selectively breeding a limited number of individuals, we stop propagating those individuals that do not possess the characteristics we desire. (2004)[1321]

When you breed dogs for a trait like largeness, nothing genetically new is made in the process. All the genetic materials that impact the size of the dogs are present at the outset, but when the materials that code for smallness are diminished through selection, the materials that code for largeness dominate. Consequently, the amount of change that can be achieved through this process is restricted. When you continue to select for a certain trait over multiple generations, you eventually hit a wall. As a professor of biology at Colorado State University explains:

> Any one breed of dogs will have less genetic variation within the breed than you find if you look across all dogs. At some point, breeding of this sort will eliminate most or all of the additive genetic variation in the breed, at least for some traits. At the point at which the additive genetic variation has been exhausted, no further "improvement" of the breed is possible even through carefully designed pairings. (2003)[1322]

Natural selection works in the same manner. It is a mechanism whereby certain genetic elements are lost because the creatures that carry them tend to die before they reproduce. As the name makes clear, natural selection is a "selection" mechanism. It affects how often genes are passed on, but has absolutely no capacity to alter the genes themselves.[1323] [1324] As explained by Ernst Mayr, the Harvard professor who authored almost 700 scientific papers and was described as "the greatest evolutionary biologist of the 20th century"; "one of the few Grand Masters of evolutionary genetics," and "the Darwin of the 20th century":[1325] [1326]

> What Darwin called natural selection is actually a process of elimination. (2001)[1327]

Textbooks almost always point to artificial selection when trying to make the case for evolution. Dogs are a favorite example, undoubtedly because the visual differences among them are more dramatic than those of any other mammal.[1328] Seldom, however, are the full implications directly stated, even though most of what is detailed above has been known for nearly one hundred years.[1329] [1330] Instead, textbooks give tacit approval to

Darwin's fable by failing to reveal that artificial and natural selection do not and cannot create new genetic materials. These processes merely involve the loss and mixing of existing genetic materials, and physical changes that take place through them are strictly limited.

MUTATIONS, THE SUPPOSED RAW MATERIAL FOR EVOLUTION

If use, disuse, artificial selection, and natural selection are incapable of creating new genetic materials, how then is evolution supposed to transpire? When genetic materials are reproduced, mistakes sometimes occur.[1331] According to evolutionists, these mistakes (called mutations) are the "raw material for evolution."[1332] We will dismantle this claim over the next several pages, and a good place to start is with the following two facts:

1) Mutations are random.[1333] While environmental factors can increase the rate at which mutations occur,[1334] they do not control or direct the nature of mutations. This same general point was made earlier. Spending a lot of time in the water will not induce you or your offspring to develop gills. We reinforce it in this particular context because some scientists have made statements that could easily lead one to believe that mutations are directed by environmental pressures.[1335] As is explained in *Principles of Genetics*, numerous experiments on mutant organisms have demonstrated that:

> environmental stress does not direct or cause genetic changes…. (2006)[1336]

For a tangible example, consider that certain *E. coli* bacteria have a mutation that makes them resistant to the antibiotic streptomycin. Experiment has proven that exposure to the antibiotic does not cause the bacteria to mutate and become resistant to it. On the contrary, the mutants are present beforehand, but when streptomycin is introduced, all of the non-mutants are killed off. Since bacteria multiply very rapidly, the survivors quickly develop into a sizeable population, all of which are resistant to streptomycin.[1337]

The fact that mutations are random and undirected by environmental factors is very disconcerting to some evolutionists because so many life forms seem to be purposely designed for the environments they inhabit. This is often expressed with the word "teleological" which means "exhibiting or relating to design or purpose especially in nature."[1338] As stated in the book, *Genetics and the Logic of Evolution* (written by a Professor and a Senior Research Scientist at Penn State University):

> There is no satisfactorily provable way out of the teleological illusion…. (2004)[1339]

Maybe that's because it is not an illusion. Even today in our advanced state of scientific knowledge, evidence of God is readily apparent, but some just arbitrarily dismiss it. Consider this statement written by Francis Crick:

> Biologists must constantly keep in mind that what they see was not designed, but rather evolved. (1988)[1340]

In other words, even though living organisms look like they were designed, we must always bear in mind that they were not. The Bible insightfully commented upon such thought processes a long time ago when it stated that people have no excuse for denying God, because although He is invisible, His existence is unmistakable from the things He has made.[1341]

2) Mutations rarely increase the fitness of an organism. The motto of evolution is "survival of the fittest," but mutations, the supposed raw materials for evolution, are far more likely to decrease the fitness of an organism than increase it. Quoting from two editions of *Principles of Genetics*:

> Most of the thousands of mutations that have been identified and studied by geneticists are deleterious and recessive. (2006)[1342]

> Mutations that increase fitness are considered to be rare because any change in a harmoniously functioning system tends to be disruptive. (1997)[1343]

> Because each gene is already the end result of a long evolutionary process, it is improbable that very many new mutations will improve a gene's function. Many mutations, like random changes in a piece of complex machinery, are likely to impair function. (2006)[1344]

The assertion that "each gene is already the end result of a long evolutionary process" is a notion we will sharply refute, but the thought that follows it conveys great insight. The term "complex machinery" was used in this statement to refer to mechanisms conceived and constructed by human beings, and just as the physical aspects of living organisms display evidence of purpose and design, so do their underlying genetic foundations. In the words of a Ph.D. computer scientist who is a firm adherent of evolution and played a major role in mapping the human genome:

> What really astounds me is the architecture of life. The system is extremely complex. It's like it was designed. (2001)[1345] [1346]

Accordingly, mutations, which are random mistakes in genetic materials, are far more likely to weaken a creature than strengthen it.

The Sole Judge of Scientific Truth Speaks Again

Evolutionists claim that mutations are the "ultimate source of all genetic variation" and that they "can produce any change that evolution has documented."[1347] [1348] In contrast, the Bible repeatedly asserts that living organisms reproduce "according to their kinds"[1349] Let's consult the experimental evidence and see which claim is more in keeping with the facts of science. For the record, the science of genetics was not founded by a group of theorists, but by a lone experimenter named Gregor Mendel.[1350] [1351] He was a Catholic monk, which means that just as the sciences of chemistry and microbiology were founded by Bible-believing experimenters, so was the science of genetics.[1352]

Certain creatures have attributes that make them ideally suited for genetic experiments. They are referred to as "model organisms." Two of the primary traits that characterize such creatures are that they can be readily monitored and have the ability to produce large numbers of offspring in short periods of time. Bacteria exemplify these qualities, and since the 1940s, extensive genetic research has been conducted upon them.[1353] [1354] [1355] Amazingly, in only 48 hours, a single bacterium can produce 10 billion offspring, which is more than the total number of people alive in the world today.[1356] [1357] Furthermore, when exposed to chemicals that increase the mutation rate, this two-day-old family of bacteria will undergo more than 400 billion mutations. This is roughly equivalent to the number of mutations that have supposedly taken place in the entire human race since the time that evolutionists claim we emerged 50,000 years ago.[1358] Remember, all this from a lone bacterium in only two days. This is why bacterial experiments have been referred to as "evolution on fast forward." (2002)[1359]

As we have already seen, mutations can cause bacteria to become resistant to antibiotics. This is often cited as proof of evolution,[1360] but a look at the underlying genetics severely undermines this claim. Consider what takes place when *E. coli* bacteria become resistant to the antibiotic streptomycin. Normally, the physical shape of streptomycin is such that it fits very precisely onto organs in the *E. coli* that manufacture proteins. When streptomycin attach to these organs, they malfunction and begin to produce defective proteins. This quickly leads to the demise of the bacterium.[1361] Experimentation has revealed there are at least 12 different mutations that can cause *E. coli* to become resistant to streptomycin.[1362] Why are there so many? Because we are dealing with a delicately balanced process that is affected by slight changes.

As is typical with drugs, the fit of the antibiotic and the bacterial organ it attaches to is so exact that even a miniscule change in the shape of either one can cause a mismatch. Picture a lock and key. If either one becomes bent or damaged, it is often difficult or impossible to insert the key into the lock or to turn it. So it is when a protein-producing organ is mutated in an area where streptomycin attaches to it. Such mutations also reduce the speed at which these organs manufacture proteins. Normally, this would be a disadvantage to the bacteria, and many of these mutants aren't even capable of surviving unless streptomycin is present, but because a slower organ makes fewer mistakes, this offsets the effect of the streptomycin and allows *E. coli* to survive in its presence. The result is not a new type of bacteria with innovative molecular machinery, but a simple variant.[1363] [1364] [1365] [1366]

Where is the supposed evolution in this process? Are mutant *E. coli* on their way to becoming dinosaurs, orange trees, porpoises, or aardvarks? The experimental evidence shows nothing that would justify such a wild claim. It takes as little as 15 minutes for an antibiotic-resistant mutant to appear in a colony of bacteria.[1367] Is there any progression observed in the hours, days, weeks, months, years, and decades that follow? Do the bacteria develop new organs or transform into other forms of life? Actually, such questions are far too trying for the theory of evolution. As Alan Linton, a creationist who headed the Department of Microbiology at the University of Bristol (England) points out:

> [T]hroughout 150 years of the science of bacteriology, there is no
> evidence that one species of bacteria has changed into another,
> in spite of the fact that populations have been exposed to potent

chemical and physical mutagens…. (2001)[1368] [1369] [1370]

What legitimate scientific basis is there to believe that bacteria have evolved into human beings when a tremendous amount of experimentation specifically designed to create evolution has failed to produce the comparatively tiny change of an *E. coli* bacterium into a *Streptococcus* bacterium? Even more damning than this for the theory of evolution are the forthcoming words written by Pierre Paul Grassé, former president of the French Academy of Sciences and the person who led 15 of France's most distinguished zoologists in producing an *Encyclopedia of the Animal World*.[1371] [1372] Beyond the fact that Dr. Grassé's statement concurs with the scientific data above, it is also consistent with the fact that bacteria sometimes mutate right back to their normal condition:[1373]

> In sum, the mutations of bacteria and viruses are merely heredi-
> tary fluctuations around a median position; a swing to the right, a
> swing to the left, but no final evolutionary effect. (1973)[1374]

Grassé didn't buy the notion that mutations were responsible for evo-lution, and he did not let the fact that he was an evolutionist stop him from saying so. He believed science would eventually reveal some other process at work.[1375] More than thirty years have passed with no such devel-opment. The problem for modern evolutionists is that they have nowhere else to turn. It has now been 150 years since Darwin published the *Origin of Species*, and of the various mechanisms that were once alleged to cause the genetic changes the theory of evolution requires, random mutation is all that is left. So despite the lack of evidence that mutations can transform one type of creature into another, evolutionists have no choice but to cling to this idea lest they be stuck in the embarrassing position of having no genetic mechanism to use in their theory.[1376] [1377]

When we assess the results of the vast array of bacterial experiments performed over the past 60 years, this simple fact emerges: bacteria remain bacteria. Moreover, they remain the same type of bacteria and do not develop novel organs. On top of this, as explained in the journal *Molecular Biology and Evolution* with regard to fossils alleged to be mil-lions of years old:

> Almost without exception, bacteria isolated from ancient material
> have proven to closely resemble modern bacteria at both morpho-
> logical and molecular levels. (2002)[1378]

If bacteria demonstrate "evolution on fast forward" and mutations are the "raw material for evolution," it should be acutely obvious to any informed and objective person how scientifically feeble this theory is.

Fly Experiments

Another creature that has been subjected to an enormous amount of genetic research is the fruit/vinegar fly, primarily a species called *Drosophila melanogaster*. These flies have been a mainstay of genetic experiments for almost a century, and the amount of research conducted on them is so enormous that a yearly conference of fruit fly researchers regularly draws more than a thousand attendees.[1379] [1380] The Encyclopædia Britannica affirms that "More data have been collected concerning the genetics of the vinegar fly than have been obtained for any other animal." (2004)[1381]

And what does this data reveal? First of all, it highlights something we discussed earlier. Selection does not give rise to new genetic materials, and consequently there are limits to the amount of change that can be achieved through selection. A classic experiment performed on fruit flies entails breeding them to maximize or minimize the number of bristles on their abdominal segments. This is done by continually mating flies that have the most or least number of bristles. When such experiments are carried out, inevitably, a limit is reached. Within 30 generations, the average number of bristles can be increased by about 60% or reduced by 30%. This, however, is the end of the line. Regardless of how many more times the flies are mated, the bristle quantities do not rise or fall beyond these limits. Continued selection produces no further response and, in fact, often results in sterility and death.[1382] Any population of flies or other living creatures carries a certain amount of genetic variability that can be exposed through selection, but selection has absolutely no capacity to add to this variability. It can only reveal that which is already present.[1383]

How about mutations? They can alter a creature's genetic composition, right? Yes, but the pertinent question is: "In what way?" There is an enormous amount of information with which to answer this question because in 1927, scientists discovered that X-rays radically increase the mutation rate in flies. Ever since then, labs from all over the world have been bombarding enormous numbers of flies with mutagens.[1384] As a result, more than 3,000 fruit fly mutations have been studied and cataloged.[1385] This massive body of research provides glaring confirmation of a point made earlier. In the words of James F. Crow, one of the most accomplished and widely cited geneticists of our time:[1386]

[T]he great bulk of [fruit fly] mutations are those that cause small, nonspecific, harmful effects." (2006)[1387]

And what of the remaining tiny fraction? Although evolutionists have tried to make their case based upon a few of these, their declarations are manifestly absurd. For instance, the one fly mutation I have seen commonly cited in support of evolution has the dramatic effect of causing an extra set of legs to grow on the fly's head where its antennae should be.[1388] The editor of *Scientific American* hails this as proof that:

> genetic mistakes can produce complex structures. (2002)[1389]

My apologies for having to state the obvious, but flies already have the genetic information for fly legs. This mutation does not produce a complex structure, but reproduces it. It's not as if the fly acquires a set of opposable thumbs that it can use to open soda bottles or a bee stinger to ward off predators. It only gets a copy of what it already has: fly legs. Furthermore, a fly that has lost its antennae (where its smell receptors reside[1390]) and has a set of legs hanging off its head hardly enjoys a survival or reproductive advantage. If such were the case, evolutionists would doubtless be touting it, and in the dozen or so writings I've consulted on this mutation, the effect on fitness is duly avoided. In sum, the mutation clearly decreases the fitness of the fly. Thus, it is an evolutionary dead end.

Enough details. Have any of the countless experiments conducted over the past one hundred years ever changed a fruit fly into a horse fly or some other type of fly? Absolutely not.[1391] [1392] Have any of these experiments produced a novel structure or something of the sort? Again, absolutely not.[1393] Given the complete lack of experimental evidence that one type of fly can turn into another or develop even a single innovative body part, how unreasonable is it to make the fanciful claim that killer whales and fruit flies are the offspring of a common ancestor?[1394]

Humans Under the Microscope

Although human beings cannot begin to compete with bacteria and insects in terms of population, we have some very sizeable numbers in comparison to many creatures such as lions, alligators, monkeys, horses, albatrosses, etc. Also, due to modern communications, it is probable that very little peculiar to the human condition escapes attention. This makes mankind a good candidate with which to study the claims of evolution.

Since evolution is about survival of the fittest, let's go straight to the crux of the matter by examining some mutations that can increase a person's fitness. Evolutionists have a favored example of just such a case, and believe it or not, it is the mutation that causes the disease of sickle cell anemia.[1395] [1396] [1397] [1398] Some background will help us understand. Red blood cells are normally round, but a mutation can cause them to take on a sickle or crescent shape as shown below.

The result of this deformation is that the cells are "fragile and easily damaged," leaving the person with a shortage of red blood cells. Symptoms include blood clots, fever, enlargement of the spleen, lack of energy, and general weakness.[1399] [1400] Before a new drug became available in the mid 1990s, the average lifespan of a person with sickle cell anemia was about 45 years.[1401] Going back to the 1960s the

Sickle cell

situation was even worse, and few with this condition lived beyond five years of age.[1402]

So, how can any of this increase fitness? When someone inherits the mutation from both of his parents, he is afflicted with the disease, but someone who inherits it from only one parent rarely experiences harmful effects.[1403] [1404] It also turns out that he is fairly resistant to malaria, a life-threatening illness that is transmitted via mosquitoes. When someone is bitten by an infected mosquito, parasites enter the bloodstream and encamp in red blood cells from which they propagate.[1405] In the case of those who carry a sickle cell mutation, this process is mitigated for reasons that are unclear.[1406] Thus, people with this mutation are less likely to contract malaria, and in areas of the world where malaria is common, we find an increased prevalence of the mutation because people who have it are more likely to survive and pass it to their children. However, there is still the potential for harm because if a mother and father both pass the mutation to their offspring, the child will have the disease.

With regards to the increased resistance to malaria this mutation provides, the primary thought that occurs to me is, "So what?" Are people who carry this mutation poised to sire descendants that are more than human? Of the multiple writings I have reviewed on this subject, I have yet to see anyone even try to make this case. The sad reality is that the only known progression of this scenario is the incurrence of sickle cell anemia. This does, however, teach us something important about evolution: The lack of evidence for it is so extreme that evolutionists are forced to rely upon impotent exemplars such as this.

It Does a Body Good

Another supposed example of human evolution is the ability of adults to digest milk. In the words of the *New York Times*:

> Throughout most of human history, the ability to digest lactose, the principal sugar of milk, has been switched off after weaning because the lactase enzyme that breaks the sugar apart is no longer needed. But when cattle were first domesticated 9,000 years ago and people later started to consume their milk as well as their meat, natural selection would have favored anyone with a mutation that kept the lactase gene switched on. (2006)

Geneticists have identified four separate mutations, each of which independently keeps the lactase gene switched on through adulthood. The *New York Times* calls this a "recent instance of human evolution."[1407]

This merits yet another yawn. Humans are mammals and are capable of digesting milk from birth. The fact that certain mutations can keep this ability active tells us nothing about where the ability came from in the first place. Furthermore, it may very well be the lactase gene was active throughout the human lifespan, but mutations partially disabled it, and what we are now seeing is other mutations or back mutations that reverse the process.

This subject also generates yet another vexing problem for those who believe in evolution. What came first? The ability of children to digest lactose or the ability of mothers to produce it? How does either of these abilities confer a survival advantage without the other? Did both just happen to evolve at the same time, in the same place, and in the same family? The bewildering intricacy of living organisms and their complex interactions with one another provide countless paradoxes of this nature that undercut the notion that humans or any other living organism are the result of blind chance acted upon by natural selection.

People are People

Of the billions of people alive today, is there anyone carrying a mutation or mutations that display the potential to transform their descendants into some other form of life? If such people exist, my search of more than 125 scientific papers, books, encyclopedias, and newspaper and magazine articles covering the topics of human genetics and evolution revealed nothing of the sort. You don't need to take my word for this, however. In the words of Pierre-Paul Grassé:

Mutations do differentiate individuals, but the human species, despite the magnitude of its population and the diversity of its habitats, both of which are conditions favorable for the evolution of the human species, exhibits anatomical and physiological stability. In wealthy western societies natural selection is thwarted by medical care, good hygiene, and abundant food, but it was not always so. Today in underdeveloped countries … natural selection can exert its pressure freely; yet the human type hardly changes. For several millennia, the Chinese have numbered hundreds of millions. The conditions of their physical and social environment have favored intensive selection. To what result? None. They simply remain Chinese. Within each population men differ by their genotype, and yet the species *Homo sapiens* has not modified its plans or structure of functions. To the common base are added a variety of diversifying and personifying ornaments, totally lacking evolutionary value. (1973)[1408]

Whether we study people, flies, or bacteria, the story is the same. The Bible's assertion that creatures reproduce "according to their kinds" is perfectly consistent with the facts of science as illuminated by experiment. This stands in stark contrast to the fanciful speculations of evolution. With regards to famous mutants such as the X-Men and Teenage Mutant Ninja Turtles, we find these characters in comic books and movies instead of walking around on the street for good reason. They aren't real.

Do People and Bananas Share a Common Ancestor or a Common Creator?

Going back at least to the time of the ancient Greeks, it has been noted that there are obvious similarities between animals.[1409] In the *Origin of Species*, Darwin proposes evolution as the explanation for this:

What can be more curious than that the hand of a man, formed for grasping, that of a mole for digging, the leg of the horse, the paddle of the porpoise, and the wing of the bat, should all be constructed on the same pattern, and should include the same bones, in the same relative positions? … The explanation is manifest on the theory of the natural selection of successive slight modifications, each modification being profitable in some way to the modified form…. (1859)[1410]

In scientific circles, this is known as the concept of "homology," a name derived from a Greek word meaning "agreement" or "same."[1411] [1412] On the surface, this seems to be an excellent point. If there are diverse creatures with clear similarities, what could be more logical than to infer that they are somehow related? However, science probes deeper than superficial traits, and when modern techniques allowed us to examine the genetic bases of such structures in the mid-1960s,[1413] gaping holes in this conceptualization became evident. As one scientist explains it, creatures don't inherit organs; they inherit genes that code for organs.[1414] Thus, if similar structures on differing creatures are the result of common ancestry, evidence for this should appear in their genes. The problem is that massive contradictions repeatedly arise when such examinations are carried out. The number and magnitude of these are so great that in 1971, Gavin de Beer, "one of the world's foremost Darwin scholars," former professor at Oxford and the University of London, past Director of the British Museum of Natural History, and recipient of the "Darwin Medal" from the Britain's national academy of science,[1415] wrote a short book entitled *Homology: An Unsolved Problem*. In it, he states:

> Because homology implies community of descent from a representative structure in a common ancestor it might be thought that genetics would provide the key to the problem of homology. This is where the worst shock of all is encountered.[1416]

He then cites specific examples to demonstrate that "characters controlled by identical genes are not necessarily" the result of common ancestry. More importantly, he also proves the converse of this to be true: some structures that appear to be the result of common ancestry actually have different genetic foundations.[1417] In other words, when you look at the genes that code for the hand of a man, a mole, the leg of a horse, the paddle of a porpoise, and the wing of a bat, the genetic underpinnings of these structures should be consistent with the notion that they derived from a common ancestor. However, in the words of de Beer:

> It is now clear that the pride with which it was assumed that the inheritance of homologous structures from a common ancestor was misplaced; for such inheritance cannot be ascribed to identity of genes. The attempt to find homologous genes, except in closely related species, has been given up on as hopeless. (1971)[1418]

The last part of this statement was premature to say the least, but the main point is still valid, as evidenced by what follows. An extensive genetic study of mammals was published in 2001. The outcome was that the majority of genetic blueprints conflict with evolutionary family trees based upon physical characteristics of these animals such as their teeth, bones, nervous system, vascular system, and other anatomical details. In the words of the study's author, some of the implications are "extremely provocative" and "suggest a radical shakeout."[1419] [1420] [1421] [1422] [1423] [1424] Another study, this one published in 2006, found in defiance of expectations that horses and bats are more genetically similar than horses and cows. Based upon this, the authors of the study proposed a new classification of mammals named after Pegasus, the flying horse of Greek mythology.[1425] [1426] Another 2006 study found that humans are genetically closer to sea slugs than we are to flies. This again was totally unexpected, and so it was surmised there must have been a common ancestor of humans, sea slugs, and flies that possessed these "surprisingly complex" genes, but the ancestors of modern-day flies lost them on the way to becoming what they are today.[1427] [1428] Then there are cases such as a 1998 study which found that animals such as sharks and llamas share a stunning genetic similarity, but since it would be outlandish to claim they are closely related, the study's authors assert that this "unusual feature" must have evolved independently.[1429] [1430] Furthermore, even within the realm of genetic/molecular analyses, various methods sometimes produce "marked inconsistencies" between them.[1431] [1432] [1433] [1434]

Is any of this starting to sound familiar? We summed up the section on the big bang with the following quote, but it is even more applicable here, especially the latter part of the closing sentence:

> Pseudoscience is often based on observations and may cite much "confirming evidence" but never permits refutation. Either the contrary data are ignored, or new details are continuously added to the theory in order to explain all new observations. (1998)[1435]

Nevertheless, when speaking of genetic similarities, some scientists make sweeping generalizations such as this:

> It is now clear that the relics of our evolution go back at least 800 million years, and that we are genetically related to all other living organisms—to a surprising degree. Bananas display a 50 per cent genetic similarity with humans, fruit flies 60 per cent, mice 65 per cent, and chimpanzees more than 98 per cent. (2005)[1436]

In actuality, the chimp figure keeps changing, and as of 2006, it was down to 94%.[1437] These disparate numbers are a result of the fact that they are all arrived at by excluding major and significant aspects of our genetic composition. As reported in the journal *Science*, there is no valid way to measure the entire extent of genetic differences between living organisms.[1438] Commenting upon the genetic disparity between humans and chimps, a geneticist at the *Max Planck Institute for Evolutionary Anthropology* states:

> I don't think there's any way to calculate a number. In the end, it's a political and social and cultural thing about how we see our differences. (2007)[1439]

In other words, such claims are based upon philosophy, not science. Furthermore, even if these figures had a legitimate basis, this concept would only make sense if a coherent picture emerged when all the pertinent data was examined. As we have seen, however, this is not the case—not by a long shot. In addition to the paradoxes detailed above and many more like them, it is revealing that evolutionary relationships constructed from genetic analyses often dramatically change when further data is included in these analyses.[1440] [1441] [1442] Such results reek of a faulty underlying paradigm, but because genetic and physical similarities are touted as evidence of evolution, there is reluctance to directly acknowledge something is fundamentally wrong. However, just as in other academic disciplines, a few individuals have the frankness to call it like they see it. Let's start with Donald R. Kaplan (University of California, Berkeley):

> No concept in biology seems to have engendered so much debate for so little gain as the concept of homology. (1984)[1443]

Alec L. Panchen (University of Newcastle on Tyne, England):

> Homology is a subject of endless fascination to biologists, but there is something about the apparently endless stream of papers and articles on the subject that suggests that the concept itself is intractable and unsatisfactory and/or that there is a lot of muddled thinking going on. (1994)[1444]

Tod F. Stuessy (Ohio State University) and Jorge V. Crisci (Museum of Natural Sciences, Argentina):

> From a general perspective, no aspect of [homology-based] analysis should be free from careful and thoughtful interpretation, and there are no easy answers or simple solutions. We have no true

[evolutionary histories], nor will we ever have them. (1984)[1445] [1446] [1447]

In view of everything above, allow me to offer a theory that is internally consistent when examined in light of all the pertinent data: Living organisms have genetic and physical similarities because they were all designed by the same being. Creations often bear the imprints of their creators. Art aficionados recognize paintings by Monet based upon their distinctive attributes. Decades later, fans of 80s rock music can identify songs from groups such as Boston because of their characteristic sound, even if they have never heard a particular song before. Like all engineers, I frequently used similar concepts, mediums, processes, and geometrical features in my designs. Creators, whether they be architects, painters, engineers, or composers, often employ an overarching methodology, the result being that they can be identified simply by examining their works. Why, then, rule out the possibility of a Creator who did exactly this, especially given the fact that the Bible states we are made in God's "image"[1448] and has repeatedly demonstrated remarkable consistency with what science has revealed about nature.

Of course, it is possible I am mistaken, and the next time you eat a banana you may want to preempt your first bite by telling it: "I'm sorry to do this to you my long-lost cousin, but I have no choice. Since all living organisms are related and I haven't yet evolved the ability to digest rocks, I am forced to consume my distant relations for nourishment." If you explain the situation in a sensible manner such as this, I am fairly confident the banana will not begrudge you.

MAGIC WANDS

One of the latest panaceas in evolution is a class of genes referred to as homeobox or Hox genes.[1449] They are extremely important in the development of organisms and often bear striking similarities in diverse creatures.[1450] Interestingly, they can function as master switches, turning on and off hundreds or thousands of other genes.[1451] In the search for information about these genes, I began by examining two popular books about evolution that use Hox genes as their genetic centerpiece. Amazingly, I failed to find even so much as an attempt to explain how these genes could have emerged or a single experiment in which they have produced a novel organ or structure.[1452] [1453] It is simply taken for granted that they appeared more than 500 million years ago in some unknown creature from which the vast majority of animals have descended.[1454] [1455] [1456] [1457]

It is amusing to watch evolutionists wave these genes around like magic wands, basically assuming that the presence of similar homeobox genes in different creatures is proof of evolution. In one of the two books mentioned above, there is a chapter entitled "From *E. coli* to Elephants" in which the discovery of "the very same" Hox genes in a wide range of creatures is greeted with cries of "Bingo" and "Jackpot."[1458] First of all, the claim is not accurate. The genes are similar, but not the "very same."[1459] Secondly, just as before, the similarities are logically explicable on the basis of a common Creator. Thirdly, the study of Hox genes has generated evolutionary paradoxes and inconsistencies like those that beset other genes, requiring repeated alterations of theory to account for new data.[1460] [1461] [1462] [1463] [1464]

Another type of wand-waving entails conveying the impression that homeobox genes can create novel body parts. In the second book referenced above, the author writes:

> [T]here are homeobox genes for eye formation and when one of them, the *Rx* gene in particular, is activated in the right place and at the right time, an individual has an eye. When something goes awry with this gene, the other homeobox genes involved in eye development cannot do their job, and the eye does not form. Clearly, the difference between having or not having an eye is a different proposition altogether from the general accretion of the bits and pieces that make up an eye. At the genetic level, major morphological novelty can indeed be accomplished in the twinkling of an eye. All that is necessary is that homeobox genes are either turned on or they are not. (1999)[1465]

Such an overview could easily create a false impression, so let's dig a little deeper. In order for a homeobox gene to trigger the development of a human eye, it must turn on hundreds or thousands of other genes operating in concert with one another to code for the many different parts of an eye. Homeobox genes don't create these other genes; they merely turn them on and off.[1466] [1467] [1468] Hence, far more is "necessary" at "the genetic level" to produce a "major morphological novelty" than homeobox genes being "turned on." Moreover, fewer than ten pages earlier in the same book, the author discusses the evolutionary "possibilities" of homeobox genes and states that "there remain important unanswered questions," one of which is:

> How will a novelty look when it does appear? (1999)[1469]

This is a clear admission that no one has ever seen a novelty appear, much less a "major" one "in the twinkling of an eye." And that is what this all boils down to. No matter how many genetic similarities we have with an elephant or hedgehog, it is the differences that must be bridged to make a compelling case for evolution, and as we have repeatedly witnessed, the evidence fails miserably on this account.

MISREPRESENTING CREATIONISTS AND PROPONENTS OF INTELLIGENT DESIGN

Faced with so many facts that conflict with their theory, some evolutionists have resorted to attacking unsourced phantom and fringe arguments that they broadly attribute to those who disagree with them. Although there are many different views held by various people who take issue with the theory of evolution, there are two clear mainstreams.[1470] The first is exemplified by organizations like Creation Ministries International, which subscribe to a literal interpretation of the Bible and can accurately be described as creationist organizations.[1471] [1472] [1473] The second is that of organizations like the Discovery Institute, which promote the concept of intelligent design. This differs from creationism in that it has "no commitment to defending Genesis, the Bible or any other sacred text." Instead, it holds that evolution may be true to a certain extent but that "certain features of the universe and of living things are best explained by an intelligent cause."[1474]

Although the positions of these organizations and the scientists who comprise them are publicly and clearly stated, they are often misrepresented by evolutionists. Such misrepresentations would never be taken seriously by anyone truly familiar with all sides of the debate, but those who are not yet informed might be dissuaded from even bothering to learn another side in the first place. For example, John Rennie, the editor of *Scientific American*, has placed the following words in the mouths of creationists:

> Mutations are essential to evolution theory, but mutations can only eliminate traits. They cannot produce new features. (2002)

He then refutes this by pointing out that mutations can make bacteria resistant to antibiotics.[1475] If such a view exists among creationists, it is far from prevalent. I've read stacks of creationist books and articles and have never seen anyone claim that mutations cannot produce traits such as antibiotic resistance.[1476] Creationists simply point out, as I have, that such mutations display absolutely no potential to transform one type of creature into another.[1477]

The next example comes from Kenneth R. Miller, Professor of Biology at Brown University, coauthor of "four enormously successful college and high school textbooks" and "one of evolution's most visible public defenders."[1478] In this chapter and the ones that follow, we are going to examine quite a bit of what Professor Miller has written. This is not to single him out for criticism, as many others have expressed similar thoughts, but to save us the time of introducing more evolutionists and reviewing their credentials. We'll start with this quote from his book, *Finding Darwin's God*:

> The opponents of evolution never deny that mutations produce variation, but they do argue that mutations, being unpredictable in their effects and random in their occurrence, cannot produce beneficial improvements for natural selection to work upon. (1999)[1479]

Once again, the example of antibiotic resistance is used to refute this statement. If what Dr. Miller affirms is true, why is it that three years prior to the publication of his book, a leading proponent of intelligent design stated precisely the opposite in an article that appeared on page A2 of the *New York Times*? These are the words of Michael Behe, who holds a Ph.D. in biochemistry and is a professor of Professor of Biological Sciences at Lehigh University:

> Mutant bacteria survive when they become resistant to antibiotics. These are all clear examples of natural selection in action. (1996)[1480]

Note that Miller and Behe are not strangers. The two of them debated in the year prior to this.[1481] Plenty of other similar quotes exist including the following from Lane Lester, a prominent creationist who holds a Ph.D. in genetics from Purdue University:

> Is there, then, no such thing as a beneficial mutation? Yes, there is. A beneficial mutation is simply one that makes it possible for its possessors to contribute more offspring to future generations than do those creatures that lack the mutation. (1998)[1482]

Dr. Lester wrote this in an article that appeared in one of the most widely read, if not the most widely read creationist publication in the world.[1483] At the time, he was the managing editor of a leading peer-reviewed creationist academic journal.[1484]

ORIGINAL SPECIES

Remarkably, Miller's claim that "the opponents of evolution" deny the existence of beneficial mutations isn't the most misleading of his comments. In the same book, he asserts:

> To the anti-evolutionist, when new species appear in the fossil record, it is not as the modified descendant of ancestral species. It must be the product of design. (1999)

This general idea appears three times in his book. In one case Miller notes there are "literally millions of insect species" and sarcastically avows that "the designer" must have been very busy making each of these.[1485] Let's commence to set the record straight by looking again at Dr. Lester's 1998 article:

> An essential feature of the creation model is the placement of considerable genetic variety in each created kind at the beginning. Only thus can we explain the possible origin of horses, donkeys, and zebras from the same kind; of lions, tigers, and leopards from the same kind; of some 118 varieties of the domestic dog, as well as jackals, wolves and coyotes from the same kind.[1486]

Let's also revisit Dr. Behe's 1996 *New York Times* article:

> I have no quarrel with the idea of common descent, and continue to think it explains similarities among species. By itself, however, common descent doesn't explain the vast differences among species.[1487]

In addition to refuting Miller's assertion, the quotes above contain some very significant points, which in order to fully grasp, requires we examine the meaning of the word "species." As explained by Ernst Mayr, who played a major role in formulating what is now the predominant definition:[1488]

> Obviously one cannot study the origin of gaps between species unless one understands what species are. But naturalists have had a terrible time trying to reach a consensus on this point. In their writings this is referred to as the "species problem." Even at present there is not yet unanimity on the definition of species. (2001)[1489]

Nevertheless, there is a definition more widely used than any other and it is this:

> An actually or potentially interbreeding population, reproductively

isolated from other such populations. (1999)[1490 1491 1492]

In other words, if creatures generally don't mate and produce fertile offspring, they are classified as different species. This does not mean they are incapable of mating and producing fertile offspring; it's just that they typically do not.[1493] As a result, there has been a tremendous proliferation of species labels. For example, in a single lake in Africa, fish known as cichlids were classified into five hundred different species because they have various colors and tend to mate within their color groups. However, the lake has become polluted and the cichlids can't see as well as they used to. Hence, they are now mating with cichlids of other colors and producing fertile offspring.[1494]

Let's make something clear. As the quotes above from Behe, Lester, other creationists,[1495 1496] and at least one forthright evolutionist demonstrate,[1497] it is not a tenet of mainstream creationism or intelligent design that these cichlids were created separately. The same applies to the 1,000 species of vinegar flies, 2,400 species of crickets, 5,000 species of ladybugs, and so forth.[1498 1499] To claim, as Professor Miller does, that intelligent design regards each species of insect as a separate act on the part of a creator is ridiculous. In fact, almost half of the North American species of crickets were identified based solely upon variations in the songs they sing.[1500] The issue is not whether crickets can give rise to crickets that sing other songs, but whether crickets arose from completely different types of creatures.

The question thus arises: Can the modification of a cricket's song or a change of color in a fish be logically extrapolated as evidence that these creatures evolved from microbes? Many evolutionists believe this,[1501 1502] but such belief requires a massive unsubstantiated leap of logic. Here the science of genetics is particularly instructive because it used to be assumed that the emergence of any new physical trait (such as color) must be the result of increased complexity. This assumption, however, has been turned on its head by the advancement of scientific knowledge. Although much of the public may be unaware of this, geneticists have known it for at least a century. In fact, William Bateson, a "dedicated Darwinist" who is credited with coining the word "genetics" and did a great deal to educate scientists about this discipline, revealed back in 1909:[1503 1504 1505 1506 1507]

> [I]t has become clear that variation, in so far as it consists in the omission of elementary factors, is the consequence of a process of "unpacking." The white Sweet Pea was created in the variation by

which one of the four color-factors was dropped out. Such variation is not, as it was formerly supposed that all variation must be,
a progress from a lower degree of complexity to a higher, but the
converse. When from a single wild type, man succeeds in producing a multitude of new varieties, we may speak of the result
as a progress in differentiation: but we must recognize that the
term is only applicable loosely, and that the obvious appearance of
increased complexity may in reality be the outcome of a process of
simplification. The facts nevertheless preclude the suggestion that
all variation even under domestication is of this nature, nor till
experimental research has developed far beyond its present limits,
can we make any confident estimate whether it is the one process
or the other which has played the larger part in the formation of
the diversity of living forms.[1508]

It has been one hundred years since this statement was published. Clearly,
experimental research has "developed far beyond" its former limits. And what
has been found? In the words of Dr. Franklin Harold, a molecular biologist
with a Ph.D. in biochemistry from the University of California and forty
years of research experience in cell biology and microbiology:[1509] [1510]

We should reject, as a matter of principle, the substitution of
intelligent design for the dialogue of chance and necessity; but
we must concede that there are presently no detailed Darwinian
accounts of the evolution of any biochemical or cellular system,
only a variety of wishful speculations. (2001)[1511]

Read that twice because the implications are astounding. If evolution
has any basis in reality, countless biochemical and cellular systems must have
evolved in order to convert microbes into oak trees, geraniums, kangaroos,
and chipmunks. Yet, despite the unbelievable amount of research and study
that has been performed in all of nature, we have yet to find even a single
example of such.

But wait. In 2006, the *New York Times* published an article with this
headline:

Study, in a First, Explains Evolution's Molecular Advance.[1512]

Here it is. An unmistakable admission by the *New York Times* that
for 145 years following the publication of the *Origin of Species*, no one
could find a single example in which evolution explained the advance of a
molecular system. As someone who reads and researches this publication

on a regular basis, I wonder why I never saw an earlier article that even hinted at this? In fact, I have read countless prior pieces in the *New York Times* in which this information was manifestly relevant and would have undercut key assertions made in those stories.[1513] Yet, this critical information was unreported.

This is not to single out one publication for intellectual dishonesty because the same can be said of almost all. With the exception of a few extraordinary instances, the major media and academia have been silent about this exceedingly important fact. Where is the forthrightness and intellectual freedom that scientists, educators, and journalists claim to champion? The truth is, as the writings of some evolutionists implicitly and explicitly reveal, there is a herd mentality and certain political bounds that cannot be breached without the risk of becoming ostracized, fired, losing funding, or being demoted.[1514 1515 1516 1517 1518]

And now that we supposedly have a lone example of evolution creating a "new piece of molecular machinery," what are we to make of it? Once the rhetoric is stripped away and the actual study results are examined, nothing exists that would justify the claim of evolution. We're not going to delve into the details here because they are convoluted, but the citations at the close of this sentence reference the academic paper on which the *New York Times* article was based and a critical assessment of it by Michael Behe.[1519 1520] If you are scientifically minded and so inclined, I encourage you to read both of these and judge for yourself which is more credible.

DARWIN'S "COLD SHUDDER"

In a chapter of the *Origin of Species* entitled "Difficulties on Theory," Darwin admits that when one examines an eye, it seems "absurd in the highest possible degree" that it "could have been formed by natural selection." However, he observes there are many degrees of complexity among the eyes of various creatures, and hence, he reasons that an eye could evolve through "numerous gradations" from very simple ("an optic nerve merely coated with pigment") to very complex ("the eye of an eagle").[1521]

When the *Origen of Species* was published, a famous scientist and confidant of Darwin's wrote to him with glowing feedback, declaring that he had "never learnt so much from one book as I have from yours," but in the same letter he wrote that "the weakest point of the book is the attempt to account for the formation of organs, the making of eyes, etc. by natural selection."

Darwin responded with appreciation for the praise and addressed the critique with these words:

> About the weak points I agree. The eye to this day gives me a cold shudder, but when I think of the fine known gradations, my reason tells me I need to conquer the cold shudder. (1860)[1522] [1523]

When Darwin wrote this, the science of genetics did not exist, and he was ignorant of the molecular complexity that existed beyond the resolution of microscopes in his era. [1524] [1525] [1526] [1527] Similarly, he was entirely unaware of the scientific facts and experiments detailed above. Thus, he saw "no great difficulty" in "believing" that an optic nerve coated with pigment could be converted into an intricate "optical instrument." [1528] Modern evolutionists do not have the same comfort of ignorance that Darwin used to conquer his cold shudder. Although the science of genetics has been in existence for more than one hundred years, we have yet to observe even a tiny step in this direction.

Nevertheless, let's put reason aside and just accept the evolutionary view of things. Let's allow that a theory requiring the formation of innumerable biochemical and cellular systems might be true, despite the failure to find a single example of such in the entire kingdom of nature.

CHAPTER 8
EMBRYOLOGY AND VESTIGIAL ORGANS

"Exceedingly Important Proofs" for the Theory of Evolution

If you graduated high school prior to about 2000, in all probability you were presented with something like this during the course of your education:

> The fact that the early development of fish, birds, and humans is similar indicates that these animals share a common ancestor.

These exact words and diagram appeared in the 1998 edition of Dr. Kenneth R. Miller's and Joseph Levine's high school biology textbook.[1529] [1530] Such "evidence" used to be a very common feature of any writing attempting to make a case for evolution.[1531] If you are not yet aware, it may astonish you to know these diagrams were derived from sketches made over a century ago and proven to be fraudulent in a 1997 study published in the *Journal of Anatomy and Embryology*.[1532] In the words of Michael K. Richardson, a staunch evolutionist and the lead author of the study:

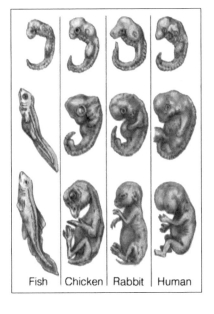

Fish | Chicken | Rabbit | Human

> This is one of the worst cases of scientific fraud. It's shocking to find that somebody one thought was a great scientist was deliberately misleading. It makes me angry. ... What he did was to take a human embryo and copy it, pretending that the salamander and

the pig and all the others looked the same at the same stage of development. They don't. ... These are fakes.[1533]

The "great scientist" referred to is Ernst Haeckel, and the story behind these drawings is one of the most sordid and fascinating in scientific history, for it vividly illustrates how preconceptions can allow intelligent and highly educated people to fall easy prey to fraud.

Although the roots of this affair predate Darwin in that other people characterized similarities between embryos, we begin our synopsis in 1859 in order to get straight to the crux of the matter. In the *Origin of Species* and in a letter written in 1860, Darwin asserted that embryology provides the most important and "strongest single class of facts in favor" of evolution.[1534] [1535] Leaning upon one of his mistaken notions of heredity, Darwin theorized that the inherited effects of "long-continued exercise" and "disuse" mainly affect animals when they are grown because that is the time in their life when they either exercise or neglect to use certain organs. Likewise, the inherited effects of use and disuse would have very limited effects on animals when they are embryos.[1536] Although Darwin felt embryology was a great source of support for this theory, he wrote letters in which he complained that none of his reviewers or friends even mentioned it. (1859/60)[1537]

Step forward a few years to 1862 when the German biologist Ernst Haeckel began to champion Darwin's theory.[1538] Over the course of several decades, he did this so forcefully and effectively that according to the *Encyclopedia of World Biography*, "in the late 19th and early 20th centuries, he was as famous as Charles Darwin."[1539] Harvard professor and famed modern evolutionist Stephen Jay Gould writes that Haeckel's books "surely exerted more influence than the works of any other scientist, including Darwin" in "convincing people about the validity of evolution." (2000)[1540] [1541] One of these books, *The History of Creation*, was considered to be "perhaps the chief source of the world's knowledge of Darwinism." (1920)[1542] In fact, Darwin, in the introduction of his own book about human evolution (*The Descent of Man*) writes of Haeckel's book:

> If this work had appeared before my essay had been written, I should probably never have completed it. Almost all the conclusions at which I have arrived I find confirmed by this naturalist, whose knowledge on many points is much fuller than mine. (1871)[1543]

In *The History of Creation*, Haeckel leaves us with a glimpse of the public attitude towards evolution at the time:

> The reproach which is now oftenest made against the Descent Theory is that it is not securely founded, not sufficiently proven. Not only its distinct opponents maintain that there is a want of satisfactory proofs, but even faint-hearted and wavering adherents declare that Darwin's hypothesis is still wanting fundamental proof. (1873)[1544]

In this book, Haeckel sets about to rectify this situation and touts a series of illustrations and remarks as "exceedingly important proofs" for the theory of evolution.[1545] The first of these is labeled "a human egg," wherein he points out that all vertebrates begin their existence as "a single cell."[1546]

Of the second, which Haeckel labels the "embryo of a mammal or bird," he writes that at "this stage of development," "all vertebrate animals in general—birds, reptiles, amphibious animals, and fishes—either cannot be distinguished from one another at all, or only by very nonessential differences."[1547] [1548]

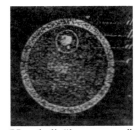

Haeckel's "human egg"

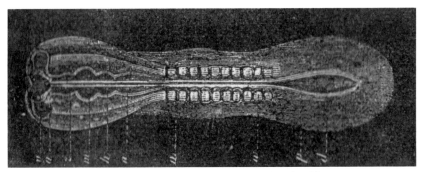

Haeckel's "embryo of a mammal or bird"

With regard to third, Haeckel claims the "facts of embryology alone would be sufficient to solve the question of man's position in nature." He asks the reader to "look attentively and compare the eight figures," and "it will be seen that the philosophical importance of embryology cannot be too highly estimated." In addition, he writes that the bones which support the gills of fish, or "gill arches," "originally exist exactly the same" in the turtle, chick, dog and human embryos shown below, and also "in all other vertebrate animals."[1549]

Tortoise	Chick	Dog	Man

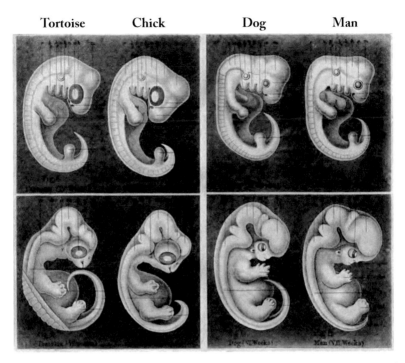

After noting that "most people" considered large-scale evolution to be "impossible," he summarizes the sketches above with these words:

> Is it not in the highest degree remarkable that all vertebrate animals of the most different classes—fishes, amphibious animals, reptiles, birds, and mammals—in the first periods of their embryonic development cannot be distinguished at all, and even much later, at a time when reptiles and birds are already distinctly different from mammals, that the dog and the man are almost identical?[1550]

Take note that Darwin echoes this claim in his next major work, referring to "the marvelous fact that the embryos of a man, dog, seal, bat, reptile, etc., can at first hardly be distinguished from each other." (1874)[1551]

After making this assertion, Haeckel appeals to these drawings to bolster a theory he had embraced a few years earlier. It entails the view that during the course of embryonic development, creatures repeat their evolutionary history. For an illustration, consider Haeckel's drawing of "a human egg", which consists of "a single cell." From this we can supposedly infer that humans evolved from amoebas or something of the sort. Also, since the embryos of turtles, chicks, dogs, and men have "gill arches," somewhere back in their evolutionary histories, they must have been fish.[1552] In other words, the physical development of a creature before it is born looks like a fast-forward movie of its evolutionary history. Scientists express this concept with the phrase, "Ontogeny [development] recapitulates phylogeny [evolutionary history]."[1553]

For future reference, I will refer to this as the theory of recapitulation. Although often ascribed to Haeckel because he did so much to popularize it, the general concept can be traced to others before him.[1554] [1555] [1556] Nonetheless, this theory became famous largely due to Haeckel's efforts, and many followed him in calling it a "law."[1557] [1558] [1559] As Dr. J. Leon Williams, a Fellow of the Anthropological Institute of Great Britain and Ireland, states in a 1913 *New York Times* article:

> It is perfectly well known that the human embryo in its development passes through the entire evolutionary process of the vertebrates.[1560]

In his next book, *The Evolution of Man* (1874), Haeckel expands his cast of embryos to include a fish, salamander, tortoise, chick, hog, calf, rabbit, and human. The result is the illustration, shown on the next page, which Haeckel boldly declares to contain more "knowledge than is afforded by the whole mass of most other sciences," and to reveal "weightier truths than are to be found in the so-called 'revelations' of all the ecclesiastical religions of the world."[1561]

"A Blind Man Could Distinguish Between Them"

In the 100+ years that followed the publication of Haeckel's book, the situation became even more sordid because, in spite of the fact that certain reputable scientists pointed out these drawings were deceptive, they were still consistently used. In 1894, the *Journal of Microscopical Science* published a paper in which the author describes his studies on the embryos of a dog-fish and a fowl. He reports the findings as such:

> There is no stage of development in which the unaided eye would fail to distinguish between them with ease…. A blind man could distinguish between them.[1563] [1564]

He footnotes this statement with a dismissive remark about Haeckel's drawings[1565] and later notes he could even tell apart the embryos of "closely allied" species such as a duck and a fowl on the second day of their development. Summarizing the matter, he writes:

But it is not necessary to emphasize further these embryonic differences; every embryologist knows that they exist and could bring forward innumerable instances of them. I need only say with regard to them that a species is distinct and distinguishable from its allies from the very earliest stages all through the development....[1566]

Fish Salamander Tortoise Chick   Hog   Calf   Rabbit   Man

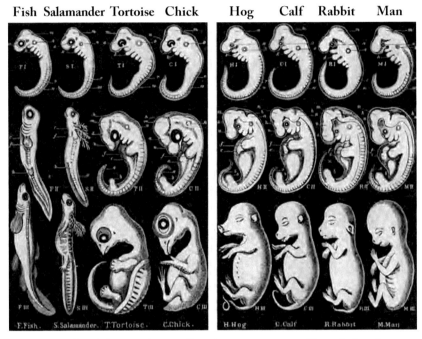

Haeckel's fraudulent embryos, alleged to contain more "knowledge than is afforded by the whole mass of most other sciences" [1562]

Why didn't the publication of this paper and the fact that embryologists were generally aware of the situation put an end to the use of Haeckel's drawings? A clue may be found in the paper itself:

I am well aware that in holding this opinion I am running counter to the great authority of Darwin.[1567]

In the early 1920s, a work entitled *The History of Biology* appeared. Originally written in Swedish, it became popular and was translated into several other languages, including English in 1929. The author was known for his use of primary sources, spending "most of his time" in a university library researching the writings of earlier scientists.[1568]

To this day, this book is "the text of choice" for the history of animal biology at the College of Science of Southern Illinois University because "no scholar has been able to write a comparable history."[1569] [1570] In this book, the author writes that Haeckel's drawings do not have "a trace of scientific value," and:

> the professional embryologists offered serious objections to them, which he either affected to overlook or else answered with personal abuse.

Yet the author defends Haeckel's character, even while noting it was difficult to reconcile this with the fact that he once displayed the exact same drawing "three times, to represent the egg of a man, an ape, and a dog." Furthermore, he states this "absurdity was removed from subsequent editions, albeit only after Haeckel had rewarded with abuse those who pointed out the fact." Given the scandalous nature of this information and the credibility of this book, why did so many modern scientists uncritically embrace Haeckel's drawings?[1571] Again, the author leaves us a clue when he summarizes the section on Haeckel by declaring that "his blunders have been forgotten, as they deserve."[1572] [1573]

This is not how science or scientists should operate. Even a child intuitively knows not to trust someone he has caught deceiving him. That is, of course, unless he deeply wants to believe what the person says. This entire affair should give pause to those who think scientists or any of us are above this weakness.

Another example that profoundly illustrates the depth of blindness created by Haeckel's drawings coupled with a selectively skeptical mindset in the academic community is scientists' apparent disregard of a textbook published in 1953. Entitled *Comparative Embryology of the Vertebrates*, this textbook actually contains a true-to-life illustration of embryo comparisons displaying a shark, rock fish, frog, chick, and human (see next page).[1574]

Judging from the resemblance of this drawing in both content and format to Haeckel's drawings, it appears the author was attempting to set the record straight. If so, he should have directly stated as such because in spite of the fact that this book constitutes the type of authoritative literature to which scientists typically appeal, they kept blindly believing in Haeckel's fraud while seemingly ignoring this illustration.

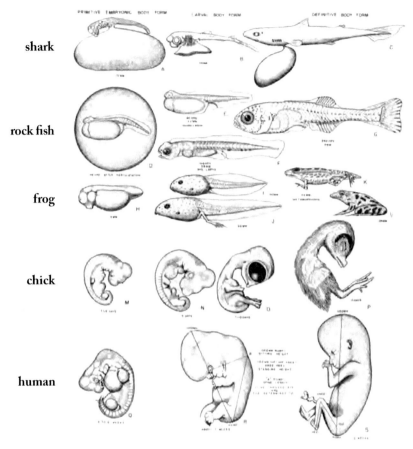

Comparative Embryology of the Vertebrates, 1953

It boggles the mind that this true depiction of embryo comparisons was published in the creditable venue that it was, and yet, 44 years later, a 1997 college textbook of embryology still reproduces a sketch derived from Haeckel's, points to the row of embryos shown below, and claims that "at an early stage all vertebrate embryos are very similar."[1575] [1576]

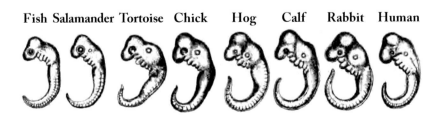

Amazingly, it gets even worse than this, for the deceptiveness of Haeckel's drawings is driven home most fully when we look at the 1997 study led by Dr. Richardson. In the 1990s, there was growing interest in a field called "evolutionary developmental biology," or "evo-devo" for short. This subject is intimately related to the study of embryos, and its popularity produced a resurgence of the idea that all or most vertebrates went through a stage in which they looked the same. The authors of the study were puzzled why "many authors" accepted this without any verifiable evidence, and set about to see if it were truly the case. To do so, they photographed and obtained pictures of 39 various vertebrate embryos at roughly the same stage of development—called the "tailbud stage"— which is the stage portrayed in the sketch above.[1577] [1578] [1579] [1580]

The paper containing the embryo photographs of the Richardson study is available online via the *Journal of Anatomy and Embryology* for a price of $34,[1581] a somewhat typical cost for scholarly journal papers. I sought permission to reproduce these photographs but have been unable to secure such permission as of the time of this writing.[1582] Thus, as a second-best alternative, the sketches below communicate the basic scientific fact of what these embryos look like. Bear in mind, however, these sketches do not fully convey the stark differences between Haeckel's drawings and reality.

Let's start with what is by far the most misleading aspect of Haeckel's sketches: the supposed similarity between "fish" and "man." As Richardson notes, Haeckel used the human embryo as his baseline and reproduced it with slight modifications, pretending all the embryos looked similar.[1583] We can see below that the drawing of "man" was fairly accurate.

Haeckel's Fish	Haeckel's Man	Actual Human

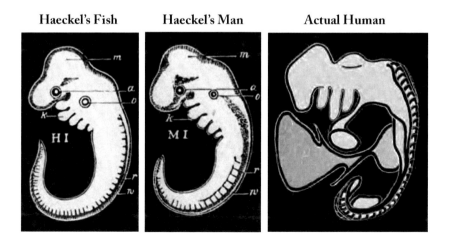

With regard to the drawing of the fish, exactly what type of fish is this supposed to be? Haeckel doesn't say. When a book of science simply labels a drawing with the word "fish," particularly in a context such as this, we get the distinct impression it is representative of all or most fish. Yet, of nine different fish examined in the Richardson study, none of them look remotely like Haeckel's. More strikingly, most of them don't even look like each other.

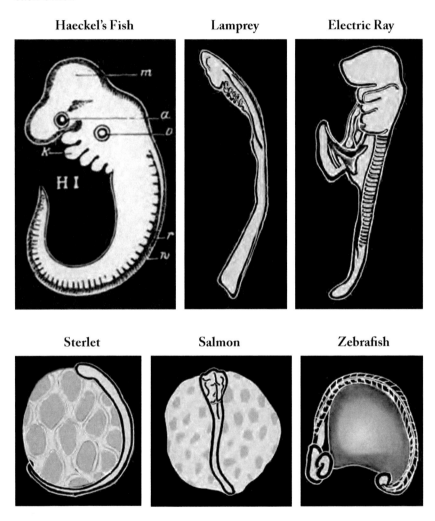

(All embryo sketches here and below are scaled to roughly the same size, but in reality, the length of the embryos in the study differ by a factor of up to ten times.[1584])

The deception does not end here. There are four other embryos in the study that correspond to Haeckel's infamous drawing. This comparison shows Haeckel's sketches next to sketches of real embryos.

Haeckel's Salamander   **Sharp-ribbed Salamander**

Haeckel's Tortoise   **European Pond Terrapin**

Haeckel's Chick

Chicken

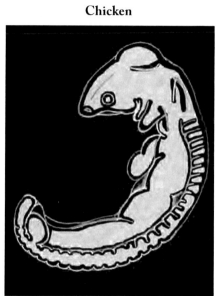

Haeckel's Rabbit

Rabbit

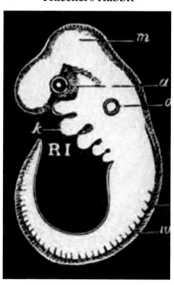

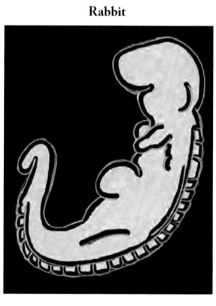

Furthermore, there are other common vertebrates that exhibit pronounced differences from the embryos above:

Domestic Cat   **Common Frog**   **Domestic Dog**

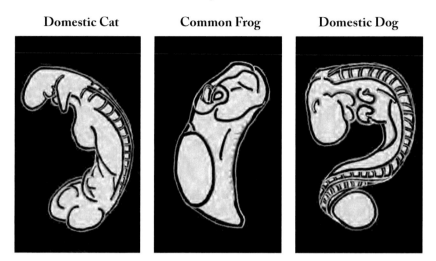

DAMAGE CONTROL

One might think this study brought a decisive end to the use of Haeckel's sketches and false claims based upon them, but censorship in the scientific community and press resulted in even more disgraceful events. In an interview with the London *Times*, Dr. Richardson revealed that at scientific conferences where he presented the study, "some Americans" voiced concern that publicizing it would "give ammunition" to creationists.[1585] So despite the fact that Haeckel's drawings had been widely touted as compelling evidence of evolution for 123 years,[1586] a search of a Lexis-Nexis database containing archives of "20,000 premium print and broadcast sources," found a total of three media outlets that ran a story about it in the period from when the study was published until a year later. These were the London *Times*, *Science*, and *New Scientist*.[1587]

For contrast, when a study about a new fossil fish alleged to be a "missing link" was published, *Time*, *Newsweek*, MSNBC, CBS, PBS, National Public Radio, the Associated Press (AP), Agence France Presse (AFP), United Press International (UPI), *New Scientist*, *USA Today*, the *Wall Street Journal*, *Boston Globe*, *Washington Post*, *Chicago Tribune*, *New York Times*, *San Francisco Chronicle*, *Philadelphia Inquirer*, *Los Angeles Times*, London *Times*, *Baltimore Sun*, and more than 90 other newspapers, wire services, magazines, and broadcast outlets published a story about it.[1588]

The result of this censorship was that many sources continued to use Haeckel's drawings and/or affirm beliefs that could only be drawn from them for years after the 1997 study exposed them as blatant frauds. These include a college neurobiology textbook (2002),[1589] two medical embryology textbooks (2003) (2004),[1590] [1591] a college textbook entitled *Evolution* (2000),[1592] *Black's Medical Dictionary* (1999),[1593] the book *What Evolution Is* by Ernst Mayr (2001),[1594] a *Time* magazine article (2002),[1595] a *National Geographic* article (2004),[1596] a popular science book published by Houghton Mifflin (2001),[1597] and a biology text written by three individuals who have collectively authored more than 160 books on science and health. (2008)[1598] [1599] If all of these writers, reviewers, and editors were unaware of Richardson's study, how could anyone expect a typical scientist or average person to be aware of it?

That's exactly what some evolutionists were hoping for. However, their worry manifested when others found the study and cited the results to contest the theory of evolution. In response, Dr. Richardson and other authors of the study sent a letter to the journal *Science*, downplaying the significance of their findings by stating, among other things, that Haeckel was simply "overzealous." (1998)[1600] Not only is this characterization irreconcilable with the pictorial evidence above, but it is also at acute odds with the language Dr. Richardson used nine months earlier: "one of the worst cases of scientific fraud," "deliberately misleading," "These are fakes." (1997)[1601]

The letter also states, "On a fundamental level, Haeckel was correct: All vertebrates develop a similar body plan (consisting of notochord, body segments, pharyngeal pouches, and so forth)." Practically speaking, this amounts to saying all vertebrates develop a spinal cord, body segments, a head, etc.[1602] [1603] [1604] One doesn't need the science of embryology to make such an observation, and this was clearly not the basis of Haeckel's or anyone else's claims relating to this matter. The assertion was that embryos show similarity far beyond what could be observed in a fully developed creature. However, as Richardson's paper emphasizes on the very first page, differences in the body plans of embryos "foreshadow important differences in adult body form."[1605] Likewise, in a paper published in 2003 on which Richardson collaborated, it is lamented that "shared features" in embryos

> are often defined so coarsely as to obscure potential variation between species. For instance, the statement that vertebrate embryos all possess a heart during the [tailbud stage] ignores

important variation in how the heart is formed….[1606]

In another attempt to limit fallout from the study, Stephen J. Gould wrote an article that appeared in the journal *Natural History*. In it, he claims to be "thoroughly mystified" how the results could be construed to challenge Darwinism. (2000)[1607] Gould focuses readers' attention on Haeckel's theory of recapitulation, asserting that "Darwinian science conclusively disproved and abandoned this idea by 1910 or so" and hardly "needs the support of a doctrine so conclusively disconfirmed from within."[1608] This statement is grossly misleading on two accounts.

First of all, although many scientists stopped viewing Haeckel's recapitulation theory as a rigid universal law, they still saw a substantial element of truth in it and propounded analogous claims throughout the 20th century up through the present era. (1949)[1609] (1956)[1610] (1985)[1611] (1988)[1612] (1988)[1613] (1999)[1614] (2001)[1615] (2009)[1616] One of the more prominent of these was the idea that embryos "repeat the embryonic stages of their ancestors," not the adult stages as Haeckel claimed. (1962)[1617] (1997)[1618] (2004)[1619] Gould undoubtedly knew about this view because it is one he expressed in the journal *Paleobiology* and in his 1977 book, revealingly entitled, *Ontogeny and Phylogeny*. (1977)[1620] [1621] It is noteworthy that Haeckel's forgeries bolster this view more than his own. They don't perpetrate a false impression of resemblance between a human embryo and an adult fish but between a human embryo and a fish embryo.[1622] [1623]

Second and more importantly, Haeckel's sketches were not just employed to support the theory of recapitulation. Regardless of whether or not evolutionists accepted this idea or some variant of it, they widely used Haeckel's drawings to support the following claim, typified in these words of Miller and Levine:

> The fact that the early development of fish, birds, and humans is similar indicates that these animals share a common ancestor. (1998)[1624]

Numerous other science books make basically the same assertion and use Haeckel's sketches to substantiate it. (1995)[1625] (1998)[1626] (1992)[1627] (1999)[1628] (1999)[1629] (1993)[1630] They have been reproduced so widely that Miller and Levine called them "the source material for diagrams of comparative embryology in nearly every biology textbook." (1997)[1631] Likewise, in August of 1999, Richardson wrote a letter to Gould stating: "I know of least fifty recent biology texts which use the drawing uncritically."[1632] Key words: "recent" and "uncritically." I have seen many examples of this, and although not each and every one explicitly affirms that embryonic similarities are evidence of common ancestry, a picture is worth a thousand words. How could anyone look at such a drawing and not consider it evidence for evolution?

Recall that by Haeckel's own admission, when his book with these sketches first appeared (1868), "most people" didn't accept evolution and even "wavering adherents" admitted it was "still wanting fundamental proof." Haeckel created "proof" in the form of these drawings, and according to *The History of Biology*, in the 1880s Darwinism achieved "victory," and "Haeckel's ideas universally prevailed without opposition."[1633] Remarking upon the two books written by Haeckel in which these drawings appeared, a prominent geneticist who was greatly impacted by the theory of recapitulation in his youth and personally met with Haeckel writes:

> The present generation can hardly understand the influence Haeckel exercised through these books upon the minds of youth, of laymen in general, and also upon large sections of the professional world. (1956)[1634] [1635]

As we have seen, this influence has continued all the way into the 21st century through the use of his drawings as evolutionary propaganda. Who knows how many minds have been swayed by this fraudulent evidence? In fact, while I was in the process of writing this book, a very intelligent and educated man told me the primary reason he accepts evolution is because "ontogeny recapitulates phylogeny."

With regard to Miller and Levine's high school biology textbook, the falsified drawings were replaced with "accurate" ones "made from detailed photomicrographs" (see picture on right).[1636] Regrettably, no matter how detailed the photos may be, the drawings lack enough clarity to make a thoughtful comparison, and they continue with Haeckel's ruse of displaying a lone sea creature labeled with

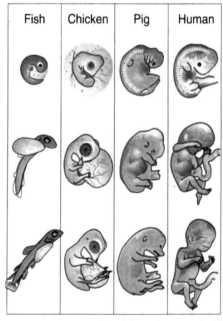

(Actual size as shown in textbook)

the word "fish."

Miller and Levine also soften their primary declaration. Instead of claiming embryonic similarities provide evidence of common ancestry, they claim, "The similarities and differences in embryonic development, which reflect the ancestry of each group of animals, provide additional evidence for evolution." (2000)[1637] In later editions, they finally get around to using a few pictures with pertinent detail but, amazingly, persist with the same basic story even though the evidence above blows holes in it. (2005)[1638 1639 1640]

You Were Aquaman

Had enough yet? I have, but unfortunately a related matter still lingers in numerous science texts. It is the demonstrably false assertion that human embryos have gills, gill slits, gill arches, gill pouches, etc. For instance, a McGraw Hill Study Guide for the Advanced Placement Biology Exam states:

> Human embryos ... actually have gills for a short time during early development, hinting at our aquatic ancestry. (2002)[1641]

How did this idea arise? Step back with me to the early 1800s, when people were pondering how a human embryo could breathe while submerged in amniotic fluid. In 1825, a German anatomist named Martin Heinrich Rathke answered this question by alleging that human embryos have gills—presumably the features in the rectangle below.[1642 1643 1644]

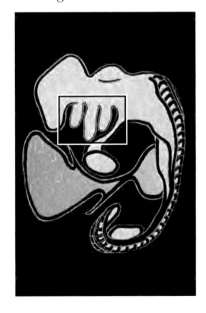

In reality, a human embryo receives oxygen from the mother's blood stream via tiny finger-like projections called chorionic villi. The oxygen is then transported to the embryo via the umbilical cord; and before the cord is developed, directly from the placenta; and before the placenta is developed, directly from the uterus.[1645 1646] Let it be noted that no one had to wait until modern times to find this out. It was demonstrated via "numerous dissections" narrated "at length" in an 1839 paper published in a French academic journal—and for scientists not fluent in the language of love, the results were summarized in *The Edinburgh Medical and Surgical Journal*—in English. Note that this synopsis directly affirmed Rathke's claim was false.[1647] [1648] Yet, once again, we will see how reality was trumped by ideology.

What exactly are the features shown above? As explained in three modern medical embryology texts, they are the precursors to various portions of our face and neck such as muscles that allow us to chew, bones that allow us to hear, our thyroid gland, and several other organs, none of which have anything to do with gills. Although many writers are still in the dark ages and use words such as "gills" or "branchial" (from the Greek word for "gills") to describe these features, the proper scientific term is "pharyngeal," from the Greek word for "throat."[1649] [1650] [1651] [1652] [1653]

It is difficult to determine exactly how much impact evolutionary beliefs played in establishing and propagating this falsity through the years, but certain facts show the role was significant and unseemly. Let's start by noting that this entire episode took place after Lamarck proposed his theory of common descent and theorized that "mammals originally came from the water." (1809)[1654] [1655] Then in 1825, Rathke avowed that human embryos have gills. Although the foundation of this claim was decisively shattered in 1839 as explained above, an extremely popular evolutionary book entitled *Vestiges of the Natural History of Creation* (1844) asserted that mammal embryos have gills.[1656] [1657] [1658] [1659] In 1850, a prominent geologist came out against this book and evolution in general, noting that microscopic examinations of these features failed to reveal even a hint of gill tissue.[1660] [1661] Later in the same decade, three eminent specialists in physiology and embryology (at least one of whom was a creationist) wrote that these features have nothing to do with gills.[1662] [1663] [1664] [1665] [1666] [1667] This, however, didn't stop Darwin from tagging them with the word "branchial" in 1859.[1668]

Moving on to Haeckel (1873), he claimed that the "gill-arches" of fish "originally exist exactly the same" in dogs, humans, and all other vertebrates. And of course, he used one of his drawings to prove this.[1669] [1670] To see is to believe, and with this image in the academic ecosystem, truth has been eclipsed in the minds of many to the present era. As is often the case with tall tales, this one exists in different forms. Occasionally, it is said human embryos actually "have gills," while other authors just claim they have "gill slits" or "gill pouches." Some authors point out that all the embryos in Haeckel's drawing have the "same number of gill arches."[1671] Gee, I wonder what the significance of that is.

A sampling of publications that have made such claims includes *National Geographic* (2007),[1672] Encyclopædia Britannica (2004),[1673] a book about evolution published by Cambridge University Press (1992),[1674] a medical reference book (1999),[1675] a high school biology textbook (1997),[1676] a college textbook on evolution (1977),[1677] a college textbook on paleontology (1988),[1678] an Advanced Placement Biology study guide (2005),[1679] a book entitled *Philosophy of Biology* (2000),[1680] a comparative embryology textbook (1949),[1681] a book entitled *Evolution and Genetics* (1962),[1682] and several other texts already mentioned above. (1999)[1683] (2001)[1684] (2001)[1685] (2002)[1686] (2002)[1687] (2004)[1688] (2008)[1689]

All of this could not fail to bring about confusion, and it has taken some very interesting forms. For example, in the college textbook *Evolution*, the author displays Haeckel's drawing, labels these features with the correct term "pharyngeal," and then mistranslates this word as "gill." (2000)[1690] Even two of the modern medical embryology texts appealed to above are still plainly under the impression of Haeckel's drawings:

> At about four weeks of development, the head and neck regions of the human embryo somewhat resemble those regions of a fish embryo at a comparable stage of development. (2003)[1691]

> Although development of pharyngeal arches, clefts and pouches resembles formation of gills in fishes and amphibia, in the human embryo real gills (branchia) are never formed. (2004)[1692]

Observe how these claims collapse when sketches of real fish and amphibian embryos are placed side by side with a human embryo (see next page).

Although these books are credible sources on the subject of human embryology, the authors obviously accepted Haeckel's sketches of other creatures as common knowledge that didn't need to be verified. And who could blame them? These sketches exist throughout academic literature. The fault for this lies not only with Haeckel but also with the scientific community at large, who has let this go on for more than a century without speaking out clearly and loudly against it. This disgraceful and lengthy episode comprising all the above is only explicable on the basis of dogmatic adherence to the theory of evolution. Academic publishers and authors would have never propagated such an easily disproved fraud if it conflicted with the theory of evolution instead of supporting it.

Nevertheless, let's overlook the fact that evolutionists widely expounded these falsities for well over a century, and some continue to do so. Let's also ignore the fact that when the deceit was exposed on several occasions, censorship and blatant attempts to misconstrue the truth of the matter followed. Let's pretend none of this undermines the credibility or objectivity of the evolutionary scientific establishment. Also, let's give the media the same benefit of a doubt. Furthermore, even though Darwin, Haeckel, Mayr, Miller, and countless others made it a point to highlight

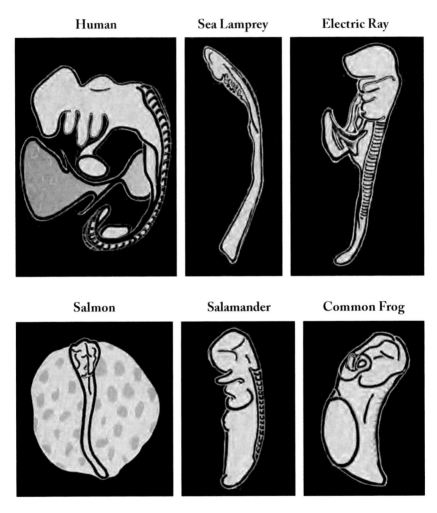

Human embryo versus embryos of fish and amphibians

diverse elements of these "exceedingly important proofs" of evolution, let's imagine the theory can get along just fine without them.

Step back for a moment and look at the overall picture of what we have witnessed thus far. Notice a pattern? The theory of evolution is built upon fables that are forced to transform with time as their foundations are knocked out one by one. However, an insular mindset will not allow some people to forgo the primary fable, and thus new false props are continually erected and shoved into place as the old ones crumble. This has been the case with the big bang theory, spontaneous generation, genetics, embryonic similarities, gills in human embryos, and, as we will see, with so-called vestigial organs and the fossil record.

VESTIGIAL ORGANS

One of the most universally cited "evidences" for evolution is the claim that humans (and other creatures) have organs that are withered relics of body parts inherited from our evolutionary ancestors. Such organs are often stamped with the label "vestigial," which is derived from the word "vestige" meaning "remnant" or "leftover." As a study guide for the Medical College Admission Test (MCAT) declares:

> Another source of evidence for evolution in comparative studies is the existence of vestigial organs. These are body parts that were presumably functional at an earlier time in an organism's evolutionary background but have since become mere remnants. Their presence can easily be accounted for by an evolutionary progression involving a gradual loss of function as lineage is modified. It would be much harder to account for nonfunctional vestiges such as the human appendix or coccyx (a remnant of tail vertebrae), if each creature were created by divine design. (1999)[1693]

Likewise, the *World Book Encyclopedia* states:

> Vestigial organs are the useless remains of organs that were once useful in an evolutionary ancestor. (2007)[1694]

Such assertions abound,[1695] [1696] [1697] [1698] [1699] [1700] echoing sentiments expressed by Lamarck in 1809 and Darwin in 1859.[1701] [1702] Once again, we will see how the theory of evolution was built upon a foundation of archaic fallacies. Although modern discoveries will constitute most of the information below, basic scientific discretion should have been enough to envelop this claim in a thick fog of skepticism right from the start. Such a scientifically rational mindset was exemplified by Pasteur, who after expressing an eminently plausible theory states:

> Should we, however, be disposed to think that such a thing must

hold true, because it seems both probable and possible, we must, before asserting our belief, recall to mind the epigraph of this work: *the greatest aberration of the mind is to believe a thing to be, because we desire it.* (1879)[1703]

Barring raw prejudice, how could anyone assert an organ is useless given all that is still unknown about the human body and nature? As reckless as it is for scientists today to make such claims, it was even more so for the likes of Lamarck and Darwin who lived in an age when the discipline of physiology was in its infancy.[1704] Even Thomas Huxley, the English biologist known as "Darwin's bulldog" for his passionate support of Darwinism, had enough discretion in 1893 to attach this footnote to an essay he had written 24 years earlier:

> The recent discovery of the important part played by the Thyroid gland should be a warning to all speculators about useless organs.[1705] [1706]

This was sound advice, especially since Haeckel had asserted the "thyroid gland is of no use whatever to man." (1874)[1707] Regrettably, many evolutionists failed to learn from this embarrassment and continued to reaffirm Darwin's claim that the human appendix is "useless." (1871)[1708] (1999)[1709] (2002)[1710] (1993)[1711] (2002)[1712] (2005)[1713] (1999)[1714] Although a good deal of the general public still labors under this fallacy, it has been discovered the appendix emits various compounds that play a role in neonatal development and in our immune system. There is still more to be learned, but from what is known thus far, it is clear the appendix not only has a function, but possibly multiple functions. Let's allow the authorities to speak for themselves, starting with the medical textbook *Surgery: Basic Science and Clinical Evidence*, a work that is the result of an ambition to "create the best, most comprehensive textbook of surgery":

> With regard to function, the widely held notion that the appendix is a vestigial organ is not consistent with the facts. Curiously, the appendix seems more highly developed in the higher primates, arguing against a vestigial role. ... Although the unique function of the appendix remains unclear, the mucosa of the appendix, like any other mucosal layer, is capable of secreting fluid, mucin, and [enzymes that break down proteins]. (2001)[1715]

Loren G. Martin, professor of physiology at Oklahoma State University writes:

For years, the appendix was credited with very little physiological function. We now know, however, that the appendix serves an important role in the fetus and in young adults. (1999)[1716]

The college textbook *Fundamentals of Anatomy & Physiology* explains the "lymphatic system" guards "against infection and disease" and:

the primary function of the appendix is as an organ of the lymphatic system. (2001)[1717]

The clinical reference book *The Human Body: an Introduction to Structure and Function*, states the appendix:

has an important function in the human specific immune system. (1999)[1718]

The biomedical research book *Medical Primatology: History, Biological Foundations and Applications* avows:

I cannot avoid mentioning the persistent mistake made by scientists who used nonprimates in research, which led to an incorrect conception about the appendix. For many decades, on the basis of using different animals, the [appendix] was considered a rudimentary organ which lost its function in humans. … On the contrary.… It is associated with the bacteriology of the intestines, the activity of the large intestine, and it plays an important role in the immunological process of the organism. (2002)[1719]

A recent paper in the *Journal of Theoretical Biology* suggests an "apparent function" of the appendix based upon five clinical findings.[1720] We'll forego the technical language and quote from a synopsis given by the Associated Press:

It produces and protects good germs for your gut. (2007)

Did the Associated Press point out that this hypothesis undermines one of the most longstanding and frequently touted claims of evidence for evolution? Of course not. Instead, the reporter found a biochemistry professor who wasn't even involved in the study to claim the theory "makes evolutionary sense."[1721]

▶One might object that if the appendix is "important," how is it that people who have theirs removed can get along fine without it? There are at least three possible reasons for this. The first is that the appendix is most active before we are born and in the earlier years of life.[1722] [1723] [1724]

Thus, removal after either of these periods has progressively less impact.

The second, quoting the 2007 paper referenced above, is that "it is anticipated that the biological function of the appendix may be observed only under conditions in which modern medical care and sanitation practices are absent."[1725] [1726]

An analogous case is that of our colon, which forms the greater part of our large intestine and plays, according to a medical textbook, "an important role" in absorbing water and sodium. Even so, it can be completely removed without any ill-effects because most people in modern societies never experience severe dehydration or sodium-depletion.[1727]

The third reason is that there may be long-term effects of appendix removal that are not yet established. Research is underway in this field, and calls for more of it have been made.[1728] For instance, removal of the appendix may be associated with increased risks of certain cancers. Back when healthy appendixes were routinely removed because most doctors thought the organ only served to make us sick,[1729] Howard R. Bierman, a clinical professor of medicine at the Loma Linda University School of Medicine in California found:

> [A]mong several hundred patients with leukemia, Hodgkin's disease, cancer of the colon and cancer of the ovaries .... 84 percent had their appendix removed years earlier. In a comparative group without cancer, 75 percent still retained the vestigial organ. (1966)[1730] ◄

UNCRITICAL THINKING

Although the appendix is the most widely cited of the so-called vestigial organs,[1731] there are many others. Darwin for instance, considered our sense of smell to be "evidence of the descent of man from some lower form":

> The sense of smell is of the highest importance to the greater number of mammals—to some ... in warning them of danger; to others ... in finding their prey.... But the sense of smell [in humans] is of extremely slight service, if any, even to the dark colored races of men.... [I]t does not warn them of danger, nor guide them to their food; nor does it prevent the [Eskimo] from sleeping in the most fetid atmosphere, nor many savages from eating half-putrid meat. (1874)[1732]

Besides the fact that science has identified various important functions for our sense of smell,[1733] why didn't Darwin consider the prospect that certain Eskimos might not have had a lot of choices as to their sleeping arrangements, and periodic meals of "half-putrid meat" may have been the only option available to some people besides starving to death? Despite such careless reasoning, Darwin is less guilty of irrationalism than many of his modern disciples because so much information is available to them that was not to him. Yet, instead of owning up to these facts, they have twisted Darwin's logic into senseless circles that epitomize the adage: "To the man with a hammer, everything looks like a nail."

Once again, history illuminates much. Parroting Lamarck,[1734] Darwin explains in the *Origen of Species* that the "main agency" in forming vestigial organs is "disuse," which "has led in successive generations to the gradual reduction of various organs, until they have become rudimentary." He ties this in with his views on embryology and everything fits together nicely, making a seemingly compelling case that he summarizes as such:

> On the view of descent with modification, we may conclude that the existence of organs in a rudimentary, imperfect, and useless condition, or quite aborted, far from presenting a strange difficulty, as they assuredly do on the ordinary doctrine of creation, might even have been anticipated, and can be accounted for by the laws of inheritance.[1735]

Critical discoveries have been made since these words were written. First of all, Darwin's principal "laws of inheritance" proved to be myths, especially those about use/disuse and natural selection being a creative process instead of an eliminative one. Yet, modern evolutionists continue to claim "vestigial" organs are incompatible with creation and can only be explained by evolution.[1736] [1737] How can it be that the same conclusion is reached when the underlying fundamentals have changed so drastically? Simple. Through uncritical thinking and circular logic. I present such examples below grouped into categories based upon their underlying fallacies. As you read these, keep in mind the overriding fallacy is the notion that vestigial organs constitute evidence for evolution. Also note that many organs fall into more than one of these categories, and throughout these examples flows the pattern of shifting stories that beset evolutionary theory. The categories are as follows:

1) Loss and degradation by mutation

2) The evolutionary scenario is incoherent

3) Circular logic/Plainly absurd

1) Loss and degradation by mutation

In the *Origen of Species*, Darwin writes the following about crabs without eyes that live in dark parts of the ocean:

> In some of the crabs the foot-stalk for the eye remains, though the eye is gone; the stand for the telescope is there, though the telescope with its glasses has been lost. As it is difficult to imagine that eyes, though useless, could be in any way injurious to animals living in darkness, I attribute their loss wholly to disuse. (1859)[1738]

To Darwin, these eyeless crabs were living proof that the theory of use/disuse held true and thus could be extrapolated to account for all sorts of evolutionary changes. Today, however, we know that even a single mutation can destroy an organ, and if its absence does not kill a creature before it reproduces or prohibit it from reproducing, the mutation can be passed along to its offspring. Blind fish and flightless winged beetles, which some evolutionists point to as supporting their stance,[1739] [1740] are simply fish and beetles with organs degraded by mutations. These creatures may be able to survive and propagate in certain environments, but they display no evidence of descent from a different type of creature any more than a person who is born blind or with a withered arm does. Even a basic knowledge of genetics shows these scenarios do not constitute evidence for the theory of evolution but exactly the opposite. They are the result of genetic degradation, consistent with the Bible's assertion that decay is a reality of nature.[1741] [1742]

The same applies to the notion that defects in nature are incompatible with the concept of intelligent design. Roughly 6% of people living in industrialized societies develop the disease of appendicitis, and studies indicate the appendix may play a role in causing inflammatory bowel disease.[1743] [1744] Hence, according to Ken Miller:

> To adopt the explanation of design, we are forced to attribute a host of flaws and imperfections to the designer. Our appendix, for example, seems to serve only to make us sick…. (1999)[1745]

Bearing in mind that the last half of this statement is discredited, the first half is based upon the absurd notion that creatures and environments do not degrade. Many evolutionists seem incapable of grasping this simple point, so I'll reiterate it from a different angle. Despite the axiom "survival of the fittest," harmful and deadly mutations continually arise and propagate in nature.[1746] [1747] [1748] [1749] [1750] On top of this, environmental factors sometimes trigger and exacerbate conditions to which such mutations predispose us.[1751] [1752] [1753] Consider the recent and dramatic increase in the number of children with peanut allergies. Genetics plays a major role in this problem, and environmental factors may also contribute.[1754] [1755] [1756] In the United States, the prevalence of this condition doubled among children between 1997 and 2002.[1757] Similarly, a study conducted in a British locality found a three-fold increase in sensitivity to peanuts between 1989 and 1996.[1758] All of this is quite compatible with the Biblical assertion that nature was originally made "good" but is now suffering from decay. These findings, however, do not accord well with the notion that evolution is making us stronger or better. The same applies to the fact that our appendix is susceptible to disease.

While we are on the subject of genetic degradation, this is a good time to address the claim made by some that the appendix is gradually disappearing in mankind.[1759] Although there is clearly nothing about the loss of an organ that weighs against creation, there is no evidence that people are born without an appendix any more frequently than what is typical for other organs. Every day in this world, people are born with heritable mutations that cause atrophied, impotent, and missing organs, and since 1718, approximately one hundred cases of people born missing an appendix have been documented.[1760] For comparison, there are roughly 200,000 people alive at this very moment who were born missing a kneecap and the same number missing a thumb.[1761] Given that a missing appendix is not evident from an external exam, determining the prevalence of this condition is not a simple matter, but according to clinical estimates, the incidence of missing thumbs, kneecaps, and appendixes is roughly the same—about 3 for every 100,000 people.[1762] [1763] In the words of a surgeon who reported on a missing appendix, this "is an exceedingly rare abnormality."[1764] [1765] [1766]

2) The evolutionary scenario is incoherent

▶Even though the longstanding myth that the appendix is useless has fallen into disrepute, some evolutionists still persist in claiming this organ provides evidence of evolution. Observe the manner in which Miller and Levine recast their argument between the 1993 and 1998 versions of their high school biology textbook:

> This appendix does not seem to serve a useful purpose today. (1993)[1767]

> The appendix is a vestigial organ that does not seem to serve a function in digestion today. (1998)[1768]

It is obvious from this that Miller and Levine got wind of the fact that the appendix does seem to serve a useful purpose. Yet, their overriding assertion remains exactly the same in both editions:

> Today, a large functioning appendix is found in some animals, such as the koala, that eat primarily plant materials. So it is probable that our appendix is left over from a time during which our ancestors needed this organ to digest their food. (1993/1998)[1769]

Note that the word "functioning" clearly implies the appendix is not functional in humans, which is misleading. Furthermore, koalas don't even have an appendix. They have an organ called a cecum, as do humans and almost every other mammal.[1770] [1771] Even Darwin recognized this fact and plainly stated it.[1772] Some background and visuals will clarify what is going on here.

The cecum of animals such as koalas and many other herbivores helps them digest what we commonly refer to as fiber or roughage. Fiber is resistant to the digestive fluids that break down proteins and carbohydrates, but given enough time, fiber can be broken down to some degree by microorganisms that reside in digestive tracts. A large cecum acts as a holding tank, allowing more time for microorganisms to break down fiber so it can be used as nutrition. A picture of a kangaroo cecum serves to illustrate this.[1773] [1774] [1775] [1776]

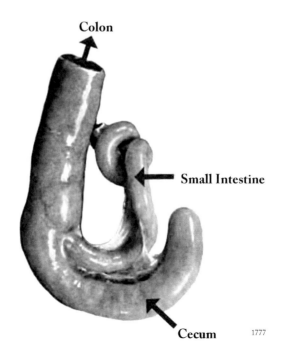

Colon

Small Intestine

Cecum   [1777]

Humans have a small cecum, and our capacity to digest fiber—especially a type of fiber known as cellulose—is comparatively limited. Cellulose is the primary ingredient of plant cell walls, and it is thought to be the most abundant organic material on earth.[1778] Items such as stalks, branches, leaves, barks, and stems are all rich in cellulose.[1779] [1780] Generally speaking, the human body is inefficient in extracting nutrition from such organic materials,[1781] [1782] [1783] which is why we cannot survive on leaves and branches. Yet, these are the mainstays of the koala's diet.[1784]

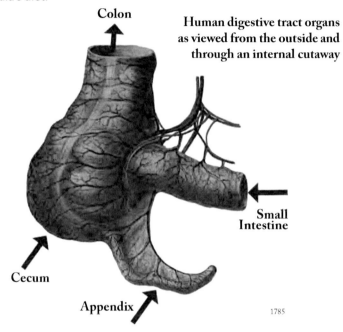

Colon

Human digestive tract organs as viewed from the outside and through an internal cutaway

Small Intestine

Cecum

Appendix

[1785]

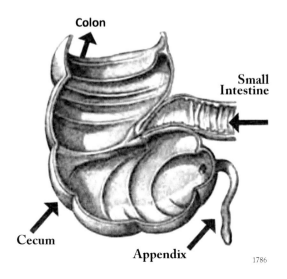

Colon

Small Intestine

Cecum

Appendix

[1786]

Looking at a human cecum, we can see the appendix is located near the end of it. The story, therefore, as Darwin, Miller and many others tell it, is that as the diet of our ancestors changed, there was no need for such a large cecum, and it shrank. Riddling this classic scenario of evolutionary adaptation with gaping holes are the following considerations and facts.

First, a comparison with other animals displays no pattern that would buttress this claim. Let's use as our baseline Miller and Levine's example of the koala, which again, has a high cellulose diet, no appendix, and a very large cecum.[1787] Supposedly, the closest living relative of the koala is the wombat,[1788] which, like the koala, has a high cellulose diet. However, it turns out the wombat has an appendix and a small cecum like those of a human.[1789] [1790] [1791] Similarly, the mountain gorilla has a high cellulose diet like that of a koala, yet it has an appendix and a small cecum like those of a human.[1792] [1793] [1794] [1795] For a stark contrast, take the case of monkeys, which are grouped into two main categories: New World and Old World.[1796] New World monkeys have a large cecum that looks very much like that of a kangaroo, and Old World monkeys have a small cecum that looks very much like that of a human. Yet, regardless of whether their cecum is large or small, neither Old World nor New World monkeys have an appendix.[1797] This has led some evolutionists to speculate that the appendix, contrary to being a vestigial organ, is a relatively new innovation.[1798] Consider also the rabbit, which has a very high cellulose diet and a large cecum like the koala. Unlike the koala, however, the rabbit has an appendix.[1799] [1800] [1801]

Such randomness in diets, supposed evolutionary relationships, cecum sizes, and the presence/absence of appendixes hardly accords with the scenario laid out by Darwin and his adherents. Of course, when dealing with a theory that presumes humans and coconut trees share a common ancestor, there is tremendous leeway for arbitrary speculation, but this hardly constitutes evidence that we are related to koalas or any other creature.

Second, even animals like koalas and rabbits with very large cecums only receive a small percentage of their nourishment from the digestive action that takes place within the cecum. Koalas can only digest about 31% of the cellulose they consume, and most of this digestion occurs in the small intestine before it even reaches the cecum. A mere 9% of koalas' total energy comes from nutrition extracted in the cecum.[1802] [1803] Moreover, rabbits that have had their cecums surgically removed show no impairment of their ability to digest cellulose.[1804]

Third, although the organs of the human intestinal tract share many similarities,[1805] there are notable anatomical, functional, and cellular differences between the appendix and the cecum.[1806] [1807] [1808] [1809] [1810] None of this bodes well for the idea that the appendix is the withered end of a cecum, but even if it were, why would anyone exclude the possibility that the human appendix is simply the withered end of a human cecum? Given that we continually see humans giving birth to humans with withered organs, what besides sheer prejudice would exclude this prospect from consideration?◄

3) Circular logic/Plainly absurd

If one blindly accepts evolution to be true, imagination can run wild with vestigial organs—and this is exactly what has happened in some cases. A 1977 college textbook written specifically for courses on evolution states the human body contains almost one hundred "useless" structures. The book only provides ten examples of such,[1811] but this is more than enough to illustrate the point to be made. In addition to the appendix, another supposedly "useless" organ is our canine teeth, yet a college text-book on human anatomy and physiology plainly refutes this assertion:

> [C]anines … are used for tearing or slashing. You might weaken a tough piece of celery by the clipping action of the incisors but then take advantage of the shearing action provided by the cuspids. (2001)[1812]

The same is true of every other such "vestigial" organ I have looked into. They clearly have a function or functions, which would explain why evolutionists have been forced to twist their logic. The prototypical argument used to be, *Look at this organ. It does nothing for us. Yet dogs have it and it works for them. The only rational explanation for this is that we evolved from a common ancestor.*[1813] Now that scientists have discovered such organs have functions and sometimes very important ones, the evolutionary story amounts to saying, *Look at this organ, Humans have it and dogs have it, but it is far more developed in dogs. It must be a vestigial organ, the presence of which proves evolution to be true.* Such arguments are entirely dependent upon the assumption that all life forms are related. The logic forms a complete circle, and by it, almost every organ in the human body but the brain could be labeled as vestigial because there is nearly always another creature with a similar organ that is stronger or more potent.

Consider the 1993 edition of Miller and Levine's high school biology text, which states: "Vestigial organs seem to serve no useful purpose at all."[1814] This sentence is stripped from the later editions of this book, as it would force evolutionists to abandon the appendix and other infamous vestigial organs like the tailbone, body hair, and so-called human "yolk sac." Examine what Miller writes about this last organ:

Our bodies do not display intelligent design so much as they reveal the evidence of evolutionary ancestry. Human embryos, for example, form a yolk sac during the early stages of development. In birds and reptiles, the very same sac surrounds a nutrient-rich yolk, from which it draws nourishment to support the growth of the embryo. Human egg cells have no comparable stores of yolk. Being placental mammals, we draw nourishment from the bodies of our mothers, but we form a yolk sac anyway, a completely empty one. (1999)[1815]

This is misleading in the extreme because a human embryo cannot survive without this organ. It is responsible for critical aspects of early blood cell production, the formation of our intestines, and the distribution of oxygen, hormones, and other substances vital to the development of the embryo.[1816 1817 1818 1819 1820] Yet, a medical book published in 2001 says it is a "semivestigial" organ that has a "more important role in other species."[1821] How much "more important" can it get than being essential for life? This goes to show that by the imaginative criteria of some evolutionists, almost any organ can be labeled as "vestigial." In reality, when it comes to the "yolk sac," "gill slits" and "branchial pouches," the only thing demonstrably vestigial are these terms. They are useless remnants from an age of ignorance about the functions of these organs. Such names would have never been applied by someone truly knowledgeable about these structures.

Just as nonsensical is the claim that male nipples somehow reveal evidence of evolution. There isn't even a need to cover the physiological details of this matter. Unless evolutionists are suddenly willing to claim male mammals evolved from female or genderless animals, this assertion is patently absurd. Yet, it appears in a 2004 *National Geographic* article.[1822] Why is it that when it comes to evolution, otherwise reasonable people don't seem to invest a moment's thought before uttering such irrational statements? My guess is that this is an emotional issue for them. It's either evolution or God, and being unwilling to accept the latter, they readily embrace nearly anything that would support the former.

There are several other highly touted "vestigial" organs, all of which suffer from factual and logical shortcomings akin to those exposed above. Rounding out this chapter are several quotes from creditable sources undermining the argument that such organs are evolutionary leftovers. We begin with wisdom teeth:

Jaws were thought to be reduced in size in the course of evolution but close examination reveals that within the species *Homo sapiens*, this may not have occurred. What was thought to be a good example of evolution in progress has been shown to be better explained otherwise. (1985)[1823] [1824]

Body hair and goose bumps (arrector pili):

> Hair keeps the body warm and serves the sensation of touch.... (1999)[1825]

> The human body has about five million hairs, and 98 percent of them are on the general body surface, not on the head. ... The five million hairs on your body have important functions. ... A **root hair plexus** of sensory nerves surrounds the base of each hair follicle. As a result, you can feel the movement of the shaft of even a single hair. This sensitivity provides an early warning system that may help prevent injury. For example, you may be able to swat a mosquito before it reaches your skin surface. (2001)[1826]

> Hairs occur in developmental-functional ... units, which comprise a hair with its follicle, sebaceous glands, and an arrector pili muscle, which may raise or lower the hair for heat conservation, sensory and communicative function, as well as for promoting the release of secretion from glands. (2005)[1827]

> ►Also, extending from the connective tissue around the hair follicle to the papillary layer of the dermis are bundles of smooth muscle, called arrector pili. ...Arrector pili muscles are controlled by the autonomic nervous system and contract during strong emotions, fear, and cold temperatures. Contraction of the arrector pili muscle erects the hair shaft, depresses the skin where it inserts, and produces a small bump on the skin surface (often called a "goose bump.") In addition, this contraction forces the sebum from the sebaceous glands onto the hair follicle and skin. Sebum oils and keeps the skin smooth, waterproofs it, prevents it from drying, and gives it some antibacterial protection. (2005)[1828] ◄

Coccyx or "tailbone":

> The coccyx provides an attachment site for a number of ligaments and for a muscle that constricts the anal opening. (2001)[1829]

> ►The narrow borders of the coccyx receive the attachment of the sacrotuberous and sacrospinous ligaments laterally, the Coccygeus

ventrally, and the Gluteus maximus dorsally. The oval surface of the base articulates with the sacrum. The rounded apex has the tendon of the Sphincter ani externus attached to it.... (1973)[1830]

Plica semilunaris or "third eyelid" (nictitating membrane):

> Structurally the nictitating membrane of Birds is a simple membranous structure containing a goodly proportion of elastic tissue (fig. 6), but that of Mammals is a complex structure. ...
>
> The physiological explanation of the plica and cartilage of Mammals is to be found in the consideration of the manner in which the eye gets rid of foreign bodies.... There can be no doubt that this is brought about by the intervention of the plica semilu-naris,—which is not so vestigial a structure as the descriptions of it might imply; it is in fact a very respectable fold, with an underlying conjunctival fornix often a quarter of an inch in depth. ...
>
> [T]he plica of man and Mammals differs entirely from the nictitating membrane of Birds. It is proposed to drop the term "nictitating membrane" in speaking of Mammals.... (1928)[1831] ◄

When all is said and done, so-called vestigial organs not only fail to supply evidence for the theory of evolution, but they also undermine it by revealing how desperate some evolutionists are to make their case in the face of evidence to the contrary. Whether it takes equivocation, circular logic, shifting stories, or outright falsities, evolutionists cannot afford to let vestigial organs go because the advances of science have already pulled the rug out from under so many of their beliefs. Nevertheless, let's give the theory the benefit of the doubt and move on to examine the fossil record.

CHAPTER 9

THE FOSSIL RECORD

According to some scientists, the fossil record provides unquestionable proof of evolution. Consider, for example, these words of George Gaylord Simpson, "the most influential palaeontologist of the mid-twentieth century":[1832]

> Fossils demonstrate that evolution is a fact. On that point there is no dissent whatever among paleontologists. Darwin had already drawn not only that conclusion from the fossil record but also added that, with improbable exceptions, all the "great leading facts in paleontology" seemed to him "simply to follow on the theory of descent with modification through natural selection." (1983)[1833]

There is a major disconnect from reality in the statement above because what Simpson refers to as "improbable exceptions," Darwin specifically calls "the most obvious and forcible of the many objections which may be urged against my theory." (1859)[1834] With regard to these objections, Oxford zoologist Mark Ridley bluntly declared:

> [T]he gradual change of fossil species has never been a part of the evidence for evolution. In the chapters on the fossil record in the *Origin of Species* Darwin showed that the record was useless for testing between evolution and special creation because it has great gaps in it. (1981)[1835]

Additionally, Ridley stated it is a:

> false idea that the fossil record provides an important part of the evidence that evolution took place. In fact, evolution is proven by a totally separate set of arguments…. (1981)

He then went on to itemize these arguments, all of which are dismantled in the previous pages of this book: homology, vestigial organs, speciation, artificial/natural selection, and cross-breeding (the mixing of existing genetic materials). Lest you imagine Ridley's statements about the fossil record could be interpreted in some other manner, he reiterated this view in the same article, writing that:

> no real evolutionist … uses the fossil record as evidence in favor of
> the theory of evolution as opposed to special creation. (1981)

I agree. The fossil record does not support the theory of evolution, and
we will start to see why by debunking an illogical notion held by some in
the general public. It has been claimed by some who haven't thought very
critically about this subject that the mere existence of dinosaur fossils and
those of other extinct creatures constitutes proof of evolution. This may
come as a shock to some people, but extinction is not evolution. Extinction is the destruction of a creature, not the production of a new one. In
fact, if creation is true, we would expect that a greater variety of creatures
lived in the past. Evolution, on the other hand, posits that all forms of life
descended from a single-celled creature, a notion that is at great tension
with the fossil record, as we will see.

TURNED TO STONE?

While we're puncturing misconceptions, let's address one with tremendous implications that bear upon much of what follows. This has to
do with the question, "What are fossils made of?" A common belief is that
fossils such as dinosaur bones do not really consist of bone matter but
minerals that have replaced it over time.[1836] [1837] [1838] In other words, many
think the organic material has long since decomposed and the bones are
totally petrified. This would make sense given the great ages cited for many
fossils, but the reality is far different. Even though minerals often seep into
bones and cause them to become heavy and brittle, far more often than
not, the original bone material is still present, even in the most ancient of
fossils. As explained by George Gaylord Simpson:

> We used to speak of fossil bone as petrified and still sometimes do
> in a loose sense. We now know, however, that it is seldom petrified
> as our ancestors understood that word and most nonpaleontolo-
> gists still do. Even in the oldest fossils the original bone substance
> has seldom "turned to rock," or even been replaced by some quite
> different material. Usually the original hard material of bone and
> teeth which formed when the animal was alive is still there. Perhaps
> it is somewhat rearranged in structure but with little or no change
> in composition, as complex compounds of lime and phosphate.
> (1953)[1839] [1840]

Citing G. B. Curry of the Department of Geology and Applied Geology, University of Glasgow:

> The fact that resistant organic molecules can survive for many millions of years has been one of the most remarkable geological discoveries of recent years…. The presence of large quantities of organic debris in rocks has long been recognized, but previously it had been widely assumed that these compounds contained no palaeontological or biological information because of the extensive degradation they had experienced. It is now clear that such an assumption is wrong, and that certain robust molecules can survive virtually intact, or at least in recognizable form, for many millions of years. (1990)[1841]

What does Curry mean by "many millions of years"? As he later points out, he means for the entire span of time living organisms with hard parts have existed.[1842] [1843] [1844] And this applies to more than just bones. As Simpson explains, coal beds supposedly tens of millions of years old contain wood "so unaltered that it can easily be cut with a saw and planed." (1953)[1845]

More striking than this is the fact that it is not just hard materials like bones, shells, and wood that survive. Bone marrow has the consistency of runny gelatin and is one of the most decay-prone tissues in the human body.[1846] [1847] [1848] [1849] Yet, bone marrow with the texture of fresh marrow and its characteristic red and yellow colors has been found in frogs and salamanders that are supposedly 10 million years old. (2006)[1850] [1851] Scientists have remarked that such spans of time are incomprehensible to the human mind,[1852] [1853] yet, in a fossil deposit claimed to be more than 40 million years old, we find insects with their original iridescent color and leaves that are still green. (1993)[1854] Another deposit alleged to be 150 million years old houses sea creatures containing ink so unaltered, that when dissolved in water, it can still be used for writing. (1990)[1855]

This is also the case at the microscopic level. When antibodies targeted for specific organic tissues were placed in contact with fossils alleged to be 70 million years old, they reacted with them, demonstrating organic materials were still present. (1990)[1856] Interestingly, even though experimentation has tentatively shown amino acids should not be able to survive for more than two million years, they have been identified in fossils claimed to be more than 440 million years old. (2001)[1857] (1990)[1858]

How is it that plants and animals normally die and decay in several years without leaving a trace of their existence,[1859] [1860] [1861] [1862] [1863] yet there exists organic materials claimed to be many millions of times older than this? Based on what is detailed above, one would be naive not to question the alleged ages of these materials, but even if they are not that old, how is it possible they could have survived for thousands or even hundreds of years?

How Fossils are Preserved

Animals and plants can be preserved in various ways like being encased in amber (as with the mosquitoes of *Jurassic Park*) or freezing (as with woolly mammoths discovered in Siberia),[1864] but the vast majority of fossils are found in sedimentary rocks.[1865] [1866] [1867] These are rocks formed from particles such as silt, sand, clay, and organic materials like seashells. Given certain conditions such as the presence of minerals that act as a cement when mixed with water, accumulations of such particles frequently bond together to form large rocks.[1868] [1869] [1870] When plants and animals are buried in sediments that later turn to stone, they become encased in what amounts to a concrete tomb. This shields them from environmental events, scavengers, and microbes that would typically bring about rapid deterioration.[1871] [1872] [1873] [1874] Yet, even in an air-tight container, organic materials continually decay, just through different means and not as quickly as they normally would.[1875] [1876] [1877]

So how do plants and animals come to be entombed in such sediments? More often than not, the answer has to do with water. Citing various scholars in the field:

> Most fossils are found in sediments which formed under water.... (1988)[1878]

> The great majority of all fossils are preserved in water-borne sediments. (1977)[1879]

> Fossils generally result when animals are buried in mud which later hardens to rock. (1956)[1880]

From this flows a very important point that is succinctly stated in the college textbook, *Evolution of Sedimentary Rocks*:

> Every area of the continents has been at one time covered by the sea, and there are some places that show clear record of being submerged at least 20 separate times. (1971)[1881]

Nothing drives this point home like the fact that even the top of Mount Everest is formed of sedimentary rock containing fossils. What type of fossils? Marine creatures such as clams.[1882] More than 300 years ago, the universal presence of marine fossils on mountain peaks and all other parts of the world led a physics professor named John Woodward to the recognition that every part of the earth has been covered by water. Given that the Bible describes a great flood (Noah's) that covered "all the high hills," Woodward interpreted the geologic record in terms of this catastrophe.[1883] Others, alternatively, sought to interpret the record in terms of a series of catastrophes.[1884]

This general view, known as catastrophism, prevailed until the late 18th and early 19th centuries when Scottish geologist James Hutton and British geologist Charles Lyell advanced the idea that most geologic features of the earth were formed in an "extremely slow" and "uniform" manner. Take special note that Darwin hailed Lyell as "My Lord High Chancellor in Natural Science" and wrote that "almost everything which I have done in science I owe to the study of his great works." (1859)[1885] (1875)[1886] Instead of explaining the geological record in terms of catastrophic events such as floods, Lyell and Hutton asserted it was better explained by "ordinary operations multiplied by time."[1887 1888 1889 1890 1891 1892] This principle, known as uniformitarianism, is described by the Encyclopædia Britannica as "the cornerstone on which the science of geology is erected." (2004)[1893] As explained in a book about fossils published by Harvard University Press, this concept has implications that bear upon the accuracy of the Bible:

> Those who accepted Lyell's book had also to accept another idea. It was common observation that rocks, which had originally been deposited beneath the sea, now formed huge cliffs, and when the thickness of *all* such sedimentary rocks from the different periods of geological time were piled on top of one another the total must be immense. Yet sediments accumulated slowly, a centimeter or two a year. So it must have taken an inconceivably long time to accumulate this great pile of sedimentary rocks; geological time must be reckoned in millions, not thousands, of years. The Biblical Creation story could not be literally true. And since fossils of many kinds were found in the sedimentary rocks, life must have been present on earth for a comparable period of time, time enough for the changes between one kind of animal and another to have happened. (1991)[1894]

On the surface this may seem logical, but the entire argument is built upon an assumption that has proven to be far from robust: "Yet sediments accumulated slowly, a centimeter or two a year." Yet again, we'll see that the sole judge of scientific truth tells us a different story.

CATASTROPHISM VERSUS UNIFORMITARIANISM

A defining feature of sedimentary rocks is the presence of many strata or layers that can be distinguished from one another,[1895] [1896] as shown in the pictures below.

© iStockphoto.com/frogdill

Cliff near Llantwit Major in South Wales, England

© iStockphoto.com/keiichihiki

Cliff at Tasman Peninsula in Tasmania, Australia

At first glance, it may appear that the only situation under which so many separate layers could form would be if they were laid down over many epochs. Consider this example of a prominent individual who operated under this supposition. Herbert Spencer was a philosopher, biologist, and contemporary of Charles Darwin, who credited Spencer as his source for the phrase "Survival of the Fittest."[1897] [1898] He is also the person who popularized the word "evolution" in the context that we now use it.[1899] Just several months before the *Origin of Species* was published, Spencer wrote an essay in which he stated that the theory of a massive flood covering our entire planet appeared to be borne out by "the most conspicuous surrounding facts," but it:

> was quite untenable if analyzed. That a universal chaotic [solvent] should deposit a series of numerous sharply-defined strata, differing from one another in composition, is incomprehensible.[1900]

As irony would have it, what Spencer considered to be "incomprehensible" has been shown to be a reality. In June of 1965, torrential rains in eastern Colorado caused flooding that destroyed buildings, roads, and bridges while depositing considerable amounts of sediments over broad areas. In the aftermath of this event, scientists decided to study an area that was relatively free of man-made structures and where a good deal sediment had accumulated. At four localities spread over a distance of about 40 miles, a total of 57 trenches were dug so that the newly formed sedimentary deposits could be examined. In every one of these of trenches, multiple sharply defined sedimentary strata were found.

The scientists who performed the study summarized their findings by calling for "special attention" to the following facts:

1. Flood plain deposits of sand several thousands of feet in width, up to at least 12 feet in thickness, and characterized by a variety of sedimentary structures may represent the accumulation of only a few hours' time.

2. Strata of sand … when deposited by a violent flood, contain dominantly horizontal layering…. Much of the layering is in the form of fine laminae [layers] similar to the type commonly ascribed to intermittent accumulation in quiet water over a long period of time. (1967)[1901]

In other words, sedimentary structures previously assumed to be the result of tranquil time-consuming processes can be formed in just hours by flooding. As explained in the academic textbook, *Flood Geomorphology*, such structures are common in flood deposits. (1988)[1902]

The strata from the 1965 Colorado flood were formed from local-ized accumulations of fairly uniform sediments. Imagine what a "univer-sal chaotic solvent" of many diverse sediments would produce. Even a simple mixture of two different types of dry sands displays an extremely unexpected phenomenon in this regard. If poured into a pile between adjacent glass plates so that we can examine the cross-section, one might guess that the different grains would sort from one another, and such is the case. Surprisingly, however, experimentation has revealed there is also a 50% likelihood that the grains will spontaneously form alternating layers as such:

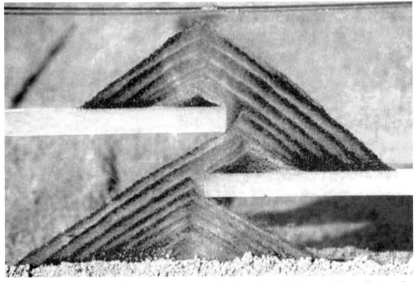

Reprinted by permission from Macmillan Publishers Ltd: *Nature*. Fineberg, Jay.
"From Cinderella's Dilemma to Rock Slides." Copyright 1997

1903 1904

Was this predicted and accounted for in previous studies of sedi-mentary rocks? As explained in the journal *Science* concerning this experiment:

Granular media show a rich variety of surprising, and at times counter-intuitive, phenomena.... (1997)[1905]

▶ Likewise, a paper in *Reviews of Modern Physics* states:

Slurries, where grains are mixed with a liquid, have a phenom-enology equally complex as the dry powders we shall describe in this article.

Thus in granular materials, shaking does not induce mixing. In contrast to ordinary liquids, where entropy favors a homogeneously mixed state, dynamics is dominant, and it leads to size separation.

Despite their apparent simplicity these materials display an intriguing range of nonlinear complex behavior, whose unraveling more often than not appears to challenge existing physics wisdom. (1996)[1906] ◄

When scientists see alternating layers like those in the diagram above, they often assume they are the result of annual patterns caused by seasonal variations in the types of sediments deposited. This is true in some instances, but as we see above and as explained by Robert Gilbert of the Department of Geography at Queens University, "many" such accounts "depend only on the inference that strong, regular cyclicity must be annual. Where the annual character cannot be established, cyclic deposits should be referred to as *rhythmites*." (2003)[1907] In plain English, people often jump to the conclusion that such layers were deposited in annual cycles with no supporting evidence. On top of this, it has been found that strata can be formed simply by the swash and backwash of waves on shorelines.[1908] There can, therefore, be little doubt that tides can have the same effect, producing multiple strata per day over vast regions.

Yet another phenomenon that rapidly produces strata is the turbulence caused by large amounts of sediments sinking in water, such as in cases of underwater landslides. Lest you suppose such circumstances are rare, according to Ben C. Kneller of the Institute for Crustal Studies at the University of California, Santa Barbara, such deposits "are amongst the most common of sedimentary deposits" and "form the largest individual sedimentary accumulations on earth." (2003)[1909]

We could examine other examples of multiple strata forming very quickly, but I think the point is sufficiently made. It should be noted that a French geologist and skeptic of Darwinism named Guy Berthault has been on the cutting edge of performing experimental work in this area. His technical papers are somewhat complex (2002)[1910], but for those interested, an online video explains his research in a way that is more easily comprehended.[1911]

NAIVE PRECONCEPTIONS

With all of the phenomena causing rapid accumulations of strata described above, a recurring theme is the presence of lateral motion. Instead of sediments sinking straight down in calm waters, there is considerable sideways movement. For obvious reasons, calm waters are not generally associated with catastrophic conditions. Of great significance, it used to be widely assumed that the clear majority of sedimentary rocks must have been deposited in "slow-moving currents or still water." This is because almost two-thirds of sedimentary rocks are made of fine sediments like silt and clay, and it was thought that such substances were too light to settle and accumulate in fast-moving waters.[1912][1913][1914]

Very recently, however, researchers have discovered that such sediments deposit "at flow velocities that are much higher than what anyone would have expected." (2007)[1915] This was not understood in the past simply because experiments did not effectively duplicate real-world conditions.[1916] The upshot, as explained in the journal *Science*, is these new findings:

> will most likely necessitate the reevaluation of the sedimentary history of large portions of the geologic record. (2007)[1917]

The same study notes that "closely spaced, parallel" strata in such rocks were widely thought to be proof of deposition in calm waters. This too, has proven to be false, and it has been determined that these strata can be formed in significant currents.[1918][1919] This has vast implications for the subject at hand, but to sum them up in the words of the geologist who led the study:

> One thing we are very certain of is that our findings will influence how geologists and paleontologists reconstruct Earth's past. (2007)[1920]

Another paper in *Science* about this study bluntly declares that "many of our preconceptions about fine-grained rocks are naive" and calls for "critical reappraisal" of all such rocks "previously interpreted as having been continuously deposited under still waters." (2007)[1921] Given past history, it will probably be untold decades or maybe never before evolutionists systematically reexamine all of the conclusions they reached while operating under these misconceptions. In the meantime, much of the scientific community labors under a 150-year-old fallacy that, because of its fine-grained-rock composition, the vast majority of the geological record must have been formed slowly over long periods of time.

Another erroneous belief many have bought into is that long periods of time are required for sediments to solidify into rock. A schoolbook entitled *The Basics of Earth Science* states that "all" sedimentary rocks are formed "under great pressures and over long periods of time." (2003)[1922] More specifically, a 1998 high school textbook says this takes millions of years,[1923] while the *World Book Encyclopedia* claims it takes thousands. (2007)[1924] Which is correct? In some cases, clearly neither. At least one type of widespread sedimentary rock has been found with items such as modern coins, coke bottle fragments, and World War II military artifacts entombed within it. (2003)[1925] (1983)[1926]

There are many types of sedimentary rocks and numerous factors that contribute to their formation, but a major, if not *the* major factor is the presence of a cementing agent. These are minerals that cause sediment grains to bond together as in concrete, which, by the way, is primarily made of two common sedimentary rocks: limestone and shale.[1927] [1928] [1929] [1930] It may not be intuitively obvious, but cementing agents are common in nature. As explained in the *Encyclopedia of Sediments and Sedimentary Rocks,* cements "include an enormous variety of minerals." (2003)[1931] Time, pressure, and temperature also play a role in the formation of sedimentary rocks,[1932] but it is important to realize there exist sediments buried more than a mile deep that are supposedly hundreds of millions of years old, and they have not turned into rock.[1933] [1934] Thus, by evolutionists' own reckoning, time and pressure are no guarantees sediments will become rock.

The fallacies and contradictions besetting the view that most of the geological record is the result of slow processes operating over vast periods of time are multitudinous. We could spend another 50 pages on this topic but instead will look into just one more area of evidence.

▶This is the discrepancy between the observed rates at which sediments accumulate, the thicknesses of sedimentary rocks, and the timeframes supposedly involved. Take the example above of the book from Harvard University Press in which it is asserted that "sediments accumulated slowly, a centimeter or two a year." (1991)[1935] [1936] Even using the lower end of this estimate, a million years time should result in about 33,000 feet or six miles of sediment.[1937] Yet, there are fossil-laden sediments allegedly deposited over one million years that are only three feet thick.[1938] This is why geologist Derek Ager has written:

> When attempts have been made to calculate rates of sedimentation in what look like continuously deposited sediments, the results look ridiculous. (1973)[1939]

In his 1993 book, Ager tones down the rhetoric a bit and calls the results "ludicrous."[1940] Is he referring to isolated instances? Absolutely not. As Ager explains, even if we account for the fact that sediments are sometimes compressed as they harden into rock:

> we are always faced with a contradiction between the rates of deposition and the known thickness of rock for a particular period of geological time. (1973)[1941]

> In fact we have an anomaly here in that the areas most commonly cited as those of continuous sedimentation without breaks … are also those of thinnest sedimentation. (1973)[1942]

Put simply, the math doesn't add up. Given that Ager accepts the theory of evolution with its accompanying long timeframes, he has no option but to assume that sedimentary rocks that appear to be continuously deposited were not.[1943] In other words, he finds himself forced to accept a scenario that conflicts with the evidence. He is caught in a similar predicament when trying to deal with the existence of 30-feet-tall upright trees buried in coal deposits. (Note that coal is generally stratified, each layer being about 1/10th of an inch thick.[1944]) He correctly observes that the trees had to be buried rapidly or they would have rotted, but when he does the simple math of dividing the thickness of the coal deposits by the alleged time it took to form them, no matter which way he turns, he ends up with results he calls "ridiculous." Hence he states:

> [W]e cannot escape the conclusion that sedimentation was at times very rapid indeed and that at other times there were long breaks in sedimentation, though it looks both uniform and continuous. (1993)[1945]

Instead of positing a conclusion at odds with the evidence, why not consider that the time scale may be mistaken? Again, these are not isolated instances. As Ager explains:

> Standing trees are known at many levels and in many parts of the world. (1993)[1946] [1947]

This, by the way, is not a new discovery. A book published in 1856 reveals:

> The coal deposits are everywhere attended with similar results. Entire trees are found, some of which are standing upright with their roots penetrating the stratum below them, exactly as they penetrated the soil on which they grew.[1948]

Here is a sketch from this book showing a mine in France with trees up to ten feet high embedded in sandstone:

1949

Let it be noted that in both the books cited above, Ager fiercely attacks creationists.[1950] [1951] Yet, as we have seen, information he brings to attention poses clear problems for the evolutionary stance. There can be little doubt he saw these implications and sought to insulate his work from being cited to contest evolution: "Situation no win."[1952] ◄

UNIFORMITARIANISM REDEFINED

As we have seen, the multiple coinciding evidences that weigh so strongly against uniformitarianism are more than sufficient to undermine this doctrine. Yet, while some outspoken evolutionists directly acknowledge this,[1953] [1954] [1955] others employ the term in a different manner, claiming that uniformitarianism is a synonym for "actualism," and it simply means the laws of nature "have not changed in the course of time." (1993)[1956] [1957] [1958]

Not only is this redefinition somewhat shallow, it is also betrayed by its historical roots and the assertions of its champion, Charles Lyell,[1959] [1960] [1961] who claimed we can "dispense with sudden, violent, and general catastrophes." (1842)[1962] Although this view is no longer tenable,[1963] [1964] the verbal shell game of redefining uniformitarianism supplies evolutionists with a graceful exit from their previous position. Just as with embryology, human gills, and vestigial organs, few had the discernment and fortitude to openly contest the academic dogma, and now that the truth has been exposed, there is an unwillingness to candidly deal with it.

WHERE DID THE WATER COME FROM AND GO?

A common criticism of the view that the Biblical flood was an actual event is that there is no physical explanation for where the waters may have come from or gone.[1965] Although this narrative is typically associated with "forty days and forty nights" of rain, the Bible also implicates tectonic activity by asserting that "all the fountains of the great deep" were "broken up."[1966] Thus, a simple prospect is that the waters came from beneath the earth's surface, and/or undersea volcanic activity displaced the ocean's waters, thereby causing them to rise. Afterwards, tectonic activity could have easily caused the waters to recede by creating highlands and basins that collected it.

This, by the way, is the basic concept evolutionists invoke to explain why the earth is not totally covered in water at this very moment. At the current rate of land erosion, it would take about 10 million years for every piece of land on earth to be worn down and washed into the oceans. Yet, evolutionists place the age of the earth at numerous multiples of this, enough to have eroded away all the land on earth many times over. How do they rationalize this? By appealing to "the constant shifting of the earth's crust." (1971)[1967] (1988)[1968]

Whatever the physical circumstances surrounding the Biblical flood may have been or not been, it bears noting that the volume of surface water on earth is roughly 15 times the volume of land above sea level. In fact, if the surface of the earth were level instead of having high and low spots, the entire globe would be submerged under about two miles of water.[1969] Hence, there can be little problem in conceiving that a flood once covered the earth. As we have seen time and again, evolutionists can be far more creative than this in envisioning scenarios to rationalize the inconsistencies that beset their theories.

Let me be clear—I am not claiming the entire geological record is explicable on the basis of a single flood. In addition to everyday processes operating over millennia, there are plenty of catastrophes that may have taken place and many others that can be shown to have taken place in recent times. For example, carbon dating tests (more to come on this later) surrounding Mt. St. Helens suggest it has erupted at least 23 times in the past 4,500 years.[1970] Thus, to recap, the numerous facts detailed above converge upon the inescapable conclusion that catastrophism has played a massive role in the formation of the geological record. Not only that, but the evidence in many cases also strongly belies uniformitarian explanations.

THE GEOLOGIC COLUMN

What do we find in the way of fossils as we dig into the earth? This is an intriguing subject, and it becomes even more so when we realize that many common notions about it are not factual. As with the rest of this book, I have translated the technical jargon into common parlance, so instead of expecting readers to know the dates assigned to geologic periods such as the Jurassic, Permian, Silurian, etc., they are supplemented with the conventional dates assigned to them using the abbreviation mya (millions of years ago). Note these dates are constantly being changed, and thus, I have used recent numbers from the *International Commission on Stratigraphy* (2008)[1971], which is the source for all such figures shown in brackets. Where the geologic period given in a quote is imprecise, such as "mid to late Cambrian," I have used the symbol ≈ to signify an approximation. Also, let it be said that I don't think these dates are accurate, so instead of placing words like "alleged" and "supposed" in front of them, realize they are implied throughout.

There is one geologic era I am going to define here for future reference. This is the Phanerozoic (Greek for "evident life"), which was originally considered to encompass the entire period life has existed. Today, it is considered to be the period in which creatures with exoskeletons have existed (from 542 million years ago to present). This era includes most of the others, with only the Precambrian preceding it. In short, it encompasses the rock strata in which the vast majority of fossils have been found.[1972] [1973] [1974] [1975] [1976] [1977]

In examining charts of geologic periods (see next page for an example), one might think this is what is found when the ground is dug at any given location. However, such charts were not made in this manner. Instead, they were made by correlating rocks and fossils from many different locations. In fact, charts more detailed than these are made by correlating rocks and fossils found at thousands of different locations.

Consider for instance the following statements of geologist Derek Ager:

> Nowhere in the world is the [stratigraphical] record, or even part of it, anywhere near complete. (1993)[1980]

> We are only kidding ourselves if we think that we have anything like a complete succession for any part of the stratigraphical column in any one place. (1973)[1981]

Or this from the college textbook, *Principles of Stratigraphy*:

The basic data of stratigraphy are derived from individual sections exposed in local areas. Synthesis becomes possible later as these local sections are correlated with one another…. (1957)[1982]

Or this from a book about evolution and palaeontology published by Cambridge University Press:

[P]reservation is uneven, periods of nondeposition and erosion exist, and embarrassing gaps preclude complete records of change…. No place on earth has a complete biological record of Phanerozoic time. (1988)[1983]

Or this from the college textbook, *The Science of Evolution*:

Probably no more than one percent of geological history can be accurately read in rocks. Furthermore, no single location contains the complete geological record. (1977)[1984]

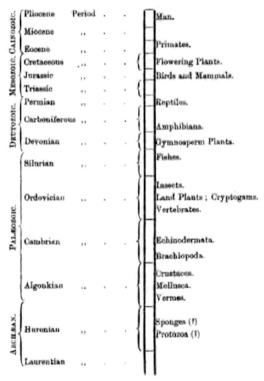

TABLE SHEWING THE RELATIVE THICKNESS OF THE ROCKS IN EACH GEOLOGICAL PERIOD.

(1902)[1978] (See this citation for another example.)[1979]

Or perhaps most enlightening, this from the college textbook, *Introduction to Geology*:

> Whatever his method of approach, the geologist must take cognizance of the following facts: (1) *There is no place on the entire earth where a complete record of the rocks is present.* ... To reconstruct the history of the earth, scattered bits of information from thousands of locations all over the world must be pieced together. The results will be at best only a very incomplete record. If the complete story of the earth is compared to an encyclopedia of thirty volumes, then we can seldom hope to find even one complete volume in a given area. Sometimes a few chapters, perhaps only a paragraph or two, will be the total geological contribution of a region; indeed, we are often reduced to studying scattered bits of information more nearly comparable to a few words or letters. (1958)[1985]

The truth to be garnered from all the above is that the geologic column which permeates the study of fossils is not the result of digging down to see what is found. Nor is it something that was assembled in a straightforward and infallible manner. As stated in a geology text published by Princeton University Press:

> For the rocks themselves are silent; the classification that orders them is something *made*. Geologists of the early nineteenth century recognized this problem themselves when they proclaimed as their task the ordering of a "chaotic" sequence of strata. ... Almost all the major divisions of the geological column were settled only after long and acrimonious debate. (1986)[1986]

So let's step back to the early nineteenth century and see what took place. Between 1815 and 1818, an English civil engineer named William Smith published several works in which he showed that the fossils and geological layers of Britain are arranged in a repeatable vertical order throughout the country. For example, in one of the lowermost of the 34 geological layers Smith identified (called "Mountain Limestone"), there are fossils of trilobites, while in the uppermost layer ("London Clay"), there are crabs. Note that all of the layers are not present in all locations (e.g., Mountain Limestone was found in 15 of 42 British counties and London Clay in 9), but the layers that appear in any particular location are in the same vertical order.[1987] [1988] [1989] [1990]

In summarizing Smith's work and the application of it by others, Dr.

Ken Miller claims the fossil record:

> told an unmistakable story—life had changed over time, changed dramatically. (1999)[1991]

Yet, in examining the fossils Smith cataloged, we do not see anything that could be reasonably described as an evolutionary progression from "lower" to "higher" forms of life. Most are shellfish and other marine creatures that sometimes look quite similar across the various layers. (If you would like to examine the fossil drawings Smith published, they are available on the website referenced in these citations.[1992] [1993])

Uppermost layer: London Clay   **18th layer: Forest Marble**

Note that Smith discovered:
- Madrepora (corals) in the 12th, 22nd, 29th, and 31st layers;
- Ostrea (oysters) in the 3rd, 5th, 11th, 12th, 14th, 15th, 16th, 19th, and 26th layers, and
- Wood in the 9th and 18th layers.

Take special note for later that corals, oysters, and wood all exist today, yet do not characterize the uppermost layers. Miller asserts that Smith used fossils to "trace geologic history."[1994] This may be Miller's interpretation of Smith's work, but Smith's writings show no evidence he adhered to such a view. As a Professor Emeritus of Geology and the History of Science at the University of New Hampshire explains, Smith used his findings for civil engineering purposes, offering "no grand theories of the earth" and no indication his findings substantiated the views of Lamarck, whose theory of evolution had been published in the prior decade.[1995] [1996] [1997] [1998] And rightfully so. As we will see, the evidence simply does not support it.

Before we continue tracing the history of the geologic column, we should recognize there is a myth among evolutionists that the geologic time scale must be correct because petroleum companies make use of it to find oil.[1999] [2000] What believers in this lore fail to realize is that the fossils employed in oil exploration are generally devoid of "chronological significance." In a book published by the American Association of Petroleum Geologists, a specialist in this field explains that "perhaps 90%" of such work is done by correlating fossils "that are not unique in time," and, it betrays a "lack of understanding" to suppose that they are. He also states that failures to grasp this are "all too numerous" and summarizes the situation in very concrete terms:

> Thus more than two dozen [strata of fossils] can be used for correlation in and around the Ventura Avenue Oil Field, yet are worthless twenty miles away. (1979)[2001]

Foundational Fallacies of the 19th Century

Now enter Charles Lyell, whose landmark book, *Principles of Geology*, was published 12 years after Smith's last major work and was "studied attentively" by Charles Darwin on his famed voyage aboard the *Beagle*.[2002] Based upon his study of the fossil record, Lyell concluded that the human race appeared on earth very recently and that:

> the extinction and creation of species has been and is the result of a slow and gradual change in the organic world. (1842)[2003]

What was the basis for these conclusions? Partly, the fact that little evidence of human remains had been found in the fossil record but primarily, Lyell's study of shellfish:

> In the present state of science, it is chiefly by the aid of shells that we are enabled to arrive at these results, for of all classes the [shellfish] are the most generally diffused in a fossil state, and may be called the medals principally employed by nature in recording the chronology of past events. (1842)[2004]

This reliance on shellfish to define geologic history "came to be widely used across Europe,"[2005] yet it represents an enormous leap of logic, akin to the one Lyell made with uniformitarianism. Contrast this rush to judgment with the attitude of Pasteur, who recognized the human tendency to "generalize by anticipation" and affirmed that scientists "should be guided by facts" and deduce from these "only such conclusions as they may strictly warrant."[2006] I'm not certain anyone can truly live up to these words, but the closer we approximate them, the less apt we are to wander the paths of ignorance while presuming we are enlightened.

This brings us to two other conclusions embraced by Lyell and his contemporaries, one quite reasonable and the other bordering on absurd. William Smith defined geological layers by the types of rocks forming them, and in his works we find layers with names that reflect this such as Chalk, Green Sand, Brickearth, Portland Rock, Forest Marble, Blue Marl, Redland Limestone, etc. Lyell, however, recognized that the consistency between the types of fossils found in each layer and the types of rock that form each layer—a consistency which generally held in England—broke down on larger scales. Studies of other areas made it impossible to accept that these relationships held on a worldwide basis. Thus, Lyell scoffed at the notion "that all the moving waters on the globe were once simultaneously charged with sediment of a red color."[2007] Yet, he accepted one very much akin to it: The belief that:

> distinct zoological and botanical provinces [i.e., local ecosystems], which form so striking a feature in the living creation, were not established at remote eras….

This is another way of saying that in each age of the past, the Earth was simultaneously covered with the same types of animals and plants. Though expressing some caution, Lyell affirmed it "seems by no means unlikely that this opinion will prove, to a certain extent, well founded…." (1842)[2008] Almost needless to say, this opinion was anything but well-founded,[2009] [2010] and worse yet, Lyell's caution about it was mostly lost upon his generation—so much so that several months before the *Origin of Species* was published, Herbert Spencer rued that this mindset permeated the academic community. In fact, he noted that "nine out of ten" geology texts gave the impression that European geological classifications were universally applicable throughout the Earth, and thus stated:

> Though, probably, no competent geologist would contend that the European classification of strata is applicable to the globe as a whole; yet most, if not all geologists, write as though it were. (1859)[2011]

Recall that this is the individual who would popularize the word "evolution" and coin the phrase "survival of the fittest." In the same essay, Spencer explained why alleged relationships between the geological systems of Europe and the U.S. were established on unconvincing evidence, denounced the concept of a uniform worldwide ecosystem as "inconsistent with the facts," and pointed out that this framework was based upon the senseless belief that:

> throughout each geologic era there has habitually existed a recognizable similarity between the groups of organic forms inhabiting all the different parts of the Earth; and that the causes which have in one part of the Earth changed the organic forms into those which characterize the next era, have simultaneously acted in all other parts of the Earth, in such ways as to produce parallel changes of their organic forms. Now this is not only a large assumption to make; but it is an assumption contrary to probability. (1859)[2012]

Now here is the crucial point. In the 40-year period between Charles Lyell's election to the London Geological Society (1819) and the publication of the statement above, ten of the twelve major divisions of the Phanerozoic column were established (i.e., Jurassic, Triassic, Cambrian, etc.).[2013] [2014] [2015] Moreover, although these divisions were pieced together in Europe, they were treated as universal throughout the world and "the original concepts survive in all their essentials" to this day. (1991)[2016] [2017] The same applies to the subdivisions that were made of these divisions.[2018] It is startling that this framework through which practically all fossil finds are interpreted was established in an era steeped in two critical fallacies: uniformitarianism and the presumption that the same animals and plant life were omnipresent over the globe in each era of the past.

Yet, this is not all. Another vital fallacy prevailed in this formative era of earth sciences—a fallacy that has only recently collapsed under the weight of reality and is still echoed in some modern texts. (1996)[2019] William Smith gave voice to it in one of his works when he asserted that fragile and well-preserved fossils could not have been washed into place by flooding without showing signs of wear because they would be damaged during transport.[2020] At first glance, this seems to make sense, but experimentation has found the exact opposite to be true. Well-preserved fossils are actually an indicator of rapid burial, which of course, is associated with flooding and other catastrophic processes. As explained by P. A. Allison at the Institute for Sedimentology, University of Reading, UK:

> A high degree of completeness of soft bodied and lightly skeletonized taxa has been used to infer minimal transport prior to

burial. ... However, a series of tumbling barrel experiments ...
have shown that this relationship does not hold....

Freshly killed organisms subject to tumbling were hardly damaged (Fig. 3A), while carcasses which had been allowed to decom-
pose for several weeks were disarticulated and fragmented....
Thus, freshly-killed organisms could tolerate turbulent transport
without fragmenting, while at the opposite extreme, carcasses
were disarticulated when buoyed up by decay gases, even in the
absence of currents. (1990)[2021]

Also, E.W. Nield (formerly of Poroperm Laboratories) and V.C.T.
Tucker (formerly of Yale Sixth Form College, U.K.) write:

[T]he possession of hard parts vastly increases an animal's chances
of being successfully fossilized. ... But even hard parts are not
indestructible, and need to be buried fairly quickly to prevent
damage. Rapid sedimentation [i.e., burial in sediments] therefore
encourages good preservation. (1985)[2022]

And likewise, R. A. Spicer of the Department of Earth Sciences at
Oxford University explains:

[B]ones and teeth can experience considerable transport without showing significant abrasion. Conversely, stationary bones
may be 'sand-blasted' by water or wind-borne sediment and thus
heavily abraded without significant transport. Thus, it is very difficult to judge the transport history of a bone from its appearance.
(1990)[2023]

Remember Lyell's use of fossil shells to inform his view of earth history?
Listen to Carl O. Dunbar (Professor of Paleontology and Stratigraphy at
Yale University) and John Rodgers (Associate Professor of Geology at Yale
University):

Abundant fossil shells likewise indicate rapid burial, for if the
shells are long exposed on the sea floor they suffer abrasion or
corrosion and are overgrown by sessile organisms or perforated by
boring animals. At the rate of deposition postulated by Schuch-
ert, 1,000 years, more or less, would have been required to bury a
shell 5 inches in diameter. With very local exceptions fossil shells
show no evidence of such long exposure. Evidently then, either
our estimates of geological time are grossly exaggerated, or else

most of the elapsed time is not represented in any given section by sedimentary deposits. (1957)[2024]

Just the opposite of what was assumed—good preservation, which represents the rule among fossil shells—indicates catastrophic burial. Consequently, fossil shells, which Lyell dubbed "the medals principally employed by nature in recording the chronology of past events,"[2025] turn out to be proof his doctrine of uniformitarianism is flawed at its core. If sediments accumulate slowly and uniformly as Lyell insisted, these shells would not be preserved in the condition we find them. Furthermore, since these shells were buried under catastrophic conditions, they may well have experienced significant transport.

"Disturbing and Significant"

Now let's begin to bring this together. While working under three simplistic and mistaken assumptions—(1) uniformitarianism, (2) the non-existence of local ecosystems in the past (even though they are everywhere today), and (3) fossils that are well-preserved cannot have experienced significant transport—it was only natural for 19th century scientists to assume fossils could be easily correlated on a global basis. Imagine if geological history consisted of eras in which the same animals and plants covered the earth, died, and were slowly buried where they fell. If well-preserved oyster fossils of the same type were found in America and Australia, it would be only natural to assume they lived in the same era. Conversely and more importantly, if different creatures were found in different strata, the same presumptions demand they must have lived in different eras.

The mindset cemented by the fallacies above is so engrained that people have had a hard time coming to grips with reality. In 1957, Dunbar and Rogers described a "discovery" in the Hall Canyon of California that they called "disturbing and significant." Here, the strata contain five distinct faunas of tiny sea creatures called foraminifers—all separated from one another in vertical order. Evolution? Ordinarily, palaeontologists would think so. However, the "disturbing" fact is that the same five faunas can be found living at different depths off the coast of California today. Thus, their position in the strata has nothing to do with evolution and everything to do with the habitat in which they lived.[2026]

Take another such example from Dunbar and Rogers; this one of a geologic formation in Illinois characterized by a repetitive sequence of ten layers, each containing a distinct fauna. In other words, if each layer is assigned a number based upon the fossils it contains, the layers are generally arranged in vertical order like this: 1-2-3-4-5-6-7-8-9-10-1-2-3-4-5-6-7-8-9-10-1-2-3-4-5-6-7-8-9-10. This phenomenon is attributed to "migrations in response to changing environment," but let's be realistic.[2027] What are the odds that the environment changed ten different ways, three times in a row, all in approximately the same order, with each resultant fauna being buried quickly enough to be preserved? A far more plausible explanation is that these faunas were subject to catastrophic burials and vertically sorted by physical factors such as habitat and buoyancy. Yet, even if we blindly accept the migration scenario, it is obvious that vertical changes in a fossil record do not necessarily constitute proof of evolution. In fact, in cases such as this, the fossil record definitively rules out evolution as an explanation for the changes.

Recall what was explained in the Princeton University Press geology text cited earlier: The "rocks themselves are silent; the classification that orders them is something *made*. Geologists of the early nineteenth century recognized this problem themselves when they proclaimed as their task the ordering of a 'chaotic' sequence of strata." How could they have done this correctly when they drew upon evidence from three critical dimensions, interpreting it backwards, upside-down, and inside-out? And how can it be that later scientists were not forced by the inconsistencies such a system would inevitability generate to abandon it? Simple. As we will see, the paradigm scientists employ in interpreting the evidence is so flexible that nearly any finding can be forced into it. In fact, the paradigm was specifically designed for this purpose.

The Guideposts of Circular Logic

Compliments of Hollywood, the word Jurassic [146–200 mya] is now inextricably linked with dinosaurs in the minds of the general public, but the origin of this word has nothing to do with dinosaurs or any other creature for that matter. Like the names of most divisions of the geologic column, it is named after a locality; in this case, the Jura Mountains of Switzerland.[2028] [2029] Likewise, the term Devonian [359–416 mya] comes from the county of Devon in England,[2030] Silurian [416–444 mya] for a people called the Silures who lived in a locality that is now part of Wales,[2031] and Cambrian [488–542 mya] from the Roman name for Wales.[2032] Such localities are referred to as "type regions," and specific geological formations in these and other locations actually define what is meant by the terms, Jurassic, Cambrian, Devonian, etc. (More details in the online citations.)[2033] [2034] [2035] However, everywhere else in the world, the fossils define the rock formation instead of vice-versa. As Dunbar and Rodgers explain:

> [C]ertain rocks in America are called the Silurian System [416–444 mya] only because of fossils contained in them…. [T]he Silurian System is in fact defined by fossils everywhere except in its type region [the place where it is defined]. And this conclusion holds for all the time-stratigraphic units. (1957)[2036]

This is not by accident. After dropping the use of rock descriptors such as "Green Sand" to identify geological periods because this practice could not be reconciled on a worldwide basis,[2037] [2038] it became standard practice to name geologic periods after selected localities. Why? The reasoning was outlined in a 1924 textbook by a professor of paleontology at Columbia University. Note that these comments pertain to smaller subdivisions of the geological column, but the same principle applies on larger scales such as the Jurassic:[2039]

> Professor Hall was one of the first in America to recognize the importance of naming formations from localities in which they were best exposed. In his report to the New York State legislature in 1839 he urges that neither [rock type] nor characteristic fossils is a satisfactory source from which to derive the name of a formation, for the first may change while the second is not always ascertainable and may even be absent. He holds it that it "becomes [essential] to distinguish rocks by names which cannot be [betrayed], and which, when the attendant circumstances are fully understood, will never prove fallacious." Such names can be derived only from localities. (1924)[2040]

In other words, 19th-century geologists intentionally established a classification system that could never be falsified, regardless of the evidence. Still, rumblings of the system's inadequacy can be found in heated arguments surrounding its implementation.[2041] As Dunbar and Rogers explain:

> Most of the bitter boundary disputes in stratigraphy have arisen because, once a boundary has been established in a standard section, stratigraphers using that standard were extremely reluctant to change it, even when it could be shown unequivocally to be incorrect relative to a standard close to the type region or to the type region itself. Commonly they have argued that the boundary

they recognized was the most natural, at least for their region, and that if a change must be made it should be made in the other standard or in the type area. In so doing, they have inevitably raised the whole question of the ultimate basis of our time-stratigraphic divisions [e.g., Jurassic, Triassic, etc.], a question that has been debated at length since the time of Werner [late 1700s] and will doubtless be debated at length for decades to come. (1957)[2042]

This is the "ultimate basis" of the so-called geologic column. A type region such as the Jurassic was defined based upon a formation in the Jura Mountains, its characteristic fossils were identified, and wherever else in the world these fossils are found, it is generally assumed the rocks are Jurassic [146–200 mya].[2043] This begs the obvious question: How were the type regions originally correlated? As explained by a Johns Hopkins University professor:

> [G]eologists of the nineteenth century divided the rock record into systems such as the Cambrian [488–542 mya], Ordovician [444–488 mya], and Silurian [416–444 mya] largely on the basis of the fossils found in the various strata. In fact, fossils continue to have primary importance in efforts to determine the relative ages of rock strata. (1993)[2044]

Hold on a moment. This is getting convoluted. Let's recap. Geologic formations throughout the world are chronologically correlated by the fossils contained in them, which were chronologically correlated by rock formations at specific localities such as the Jura Mountains, which were chronologically correlated by fossils contained in them. Once again, an obvious question arises: How were these fossils originally correlated? If you are thinking radioactive dating, wrong answer. The geological column was established long before the development of radioactive dating,[2045] [2046] and as Nield and Tucker explain, the "accuracy" of radioactive dating is "very poor indeed, and it is never used in the correlation of sediments." (1985)[2047] The truth is these fossils were correlated by interpreting the geological and fossil records as they were understood in the 1800s based upon all the fallacies detailed above and those yet to follow.

   Some evolutionists have cautiously admitted to the obvious truth that circular reasoning is involved in the construction of the geologic column but justify it by rationalizing it is a "process of trial and error."[2048] [2049] However, the practical reality is that any discovery, even one that is a "complete shock," can be forced into the geologic column without testing the system's basis in reality. Prepare for what follows.

WHAT DID DINOSAURS EAT?

A long-standing feature of museum displays that depict dinosaurs in their native habitats is the absence of grass. Instead, foliage such as ferns adorn the landscape.[2050] Why? An article in the magazine *New Scientist* explains:

> Textbooks have long taught that grasses did not become common until long after the dinosaurs died at the end of the Cretaceous period, 65 million years ago. Depicting dinosaurs munching on grass was considered by experts to be as foolish as showing prehistoric humans hunting dinosaurs with spears. (2005)[2051] [2052] [2053]

In his 1999 book, *Finding Darwin's God*, Professor Ken Miller mockingly offers young-earth creationists "a stunning opportunity to validate their ideas." "All they would have to do," he wrote, was "pick through" fossilized dinosaur dung and:

> find a single contemporary organism. Seeds or microscopic pollen grains from modern plants would do the trick in the case of herbivorous dinosaurs.[2054]

Besides the fact that all grass pollens look the same and there is no known way to distinguish contemporary grass pollen from that of any alleged ancestor,[2055] if a creationist ever uncovered such evidence it would be ignored or dismissed by Dr. Miller and his ilk. Thus, as providence would have it, an evolutionist unveiled it instead. In 2005, the journal *Science* published research that a scientist at the Swedish Museum of Natural History referred to as a "complete shock."[2056] In fossilized dinosaur feces (Isn't the irony almost unbearable?), remains of five different grasses were discovered, all of them consistent with living grasses.[2057] [2058] [2059] Furthermore, these remains correspond to a range of grass types considered to be among the "more highly evolved" of modern grasses.[2060] As explained in a paper accompanying the original research:

> These remarkable results will force reconsideration of many long-standing assumptions about grass evolution, dinosaurian ecology, and early plant-herbivore interactions.[2061] (2005)

Now that Dr. Miller's "stunning opportunity" for creationists to "validate their ideas" is a reality, can we expect a change in the scientific consensus? Don't hold your breath because, as explained earlier, the evolutionary paradigm is so pliable that no matter what is found, the theory can be altered to make it fit after the fact. This is a wonderful attribute if you are a partisan supporter of evolution. If, however, your objective is to practice science, a theory that is not falsifiable is not scientific. As explained in the college textbook *Foundations of Modern Cosmology*:

> Falsifiability unambiguously distinguishes scientific from nonscientific explanations. (1998)[2062] [2063]

Furthermore, the unfalsifiable nature of evolutionary palaeontology extends beyond its infinite flexibility. It also avoids contending with the sole judge of scientific truth. As an Oxford zoologist wrote:

> Palaeontology, certainly as much as any other branch of biology, and perhaps more than most, is prone to speculation. This consists of ideas that cannot be falsified, because suitable methods for testing them are simply not available. (1982)[2064]

Understanding this, let's examine some other cases where evolutionary interpretations of the fossil record bear these marks of nonscientific explanations.

"Unbelievable" Similarity

The peculiar-looking crustaceans depicted below are called mussel-shrimps, also known as seed shrimps, ostracodes, or ostracods. These creatures are typically a fraction of an inch long and can be found in oceans, lakes, rivers, and ponds. They are also abundant in the fossil record, and thousands of species have been labeled, both living and extinct.[2065] [2066] [2067] [2068]

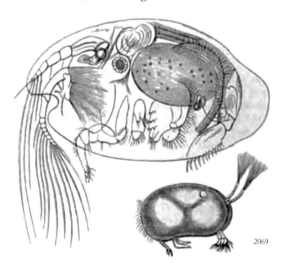

2069

In 2003, a paper in the journal *Science* analyzed a mussel-shrimp fossil discovered in rocks classified as Lower Silurian—or approximately 425 million years old. Besides enjoying the distinction of allegedly having the world's oldest fossilized penis, this creature has another noteworthy feature. With one minor exception, it looks "almost exactly" like a living mussel-shrimp all the way "down to two hairs on the end of its swimming appendages." Listen to scientists remarking upon this discovery: "I was flabbergasted"—"unbelievable to see the similarity with the living forms."[2070] [2071] [2072] [2073]

"Unbelievable" is an apt choice of word. To get a feel for what 425 million years ago looks like in the theory of evolution, this is well before there supposedly lived a creature whose direct descendants include humans, mice, cows, kangaroos, chickens, newts, and carps.[2074] The next time you see a newt, consider that evolutionists would have you believe it is your blood relative and that your mutual great, great, great… grandmother lived 75,000,000 years after a mussel-shrimp that looks almost exactly like those alive right now.[2075]

Once again, no matter what fossil evidence is discovered and how outlandish its implications are for evolutionary theory, evolutionists dutifully force it into their framework without thinking to question the legitimacy of that framework. It's like the old Meineke muffler commercial in which a brand X muffler shop mechanic is about to install an enormous muffler into a compact car and the customer asks, "Are you sure it will fit?" The mechanic's response is, "Don't worry. We'll make it fit." As we have seen and shall see, this mindset permeates the core of evolutionary thinking.

Index Fossils

Some fossils play an eminent role in evolutionary theory. These are referred to as index, guide, or zone fossils. As explained in a college textbook entitled *The Science of Evolution*:

> Certain fossils appear to be restricted to rocks of a relatively limited geological age span. These are called *index fossils*. Whenever a rock is found bearing such a fossil, its approximate age is automatically established. (1977)[2076]

The academic text, *Principles of Paleontology*, states the same more technically:

> [I]ndex, guide or zone fossils have been used widely in the correlation of Phanerozoic strata, and global biostratigraphic intervals have often been recognized on the basis of one or more pervasive,

diagnostic fossil species. (2007)[2077]

So why are "Jurassic" fossils found in "Jurassic" rocks? Simple, because Jurassic rocks are "automatically" labeled as such based upon the fossils found in them. This amounts to a massive incestuous circular of logic—one that you might think could never be broken under any circumstances. Yet, it has been broken in ways that are unmistakable. *The Science of Evolution*, continuing in its explanation of index fossils, explains:

> This method is not foolproof. Occasionally an organism, previously thought to be extinct, is found to be extant. Such "living fossils" obviously cannot function as index fossils except within the broader time span of their known existence. (1977)[2078]

Let me be a little more direct. Creatures alleged to have gone extinct millions of years ago have been found alive. Yes, that means living among us at this very moment.

The Lazarus Effect

In early 2005, scientists touring a food market in Laos came across an item for sale later dubbed the "Laotian rock rat." Since it was a unique-looking creature with a head like a rat and tail like a squirrel,[2079] they purchased a few and sent them for study to London's Natural History Museum.[2080] The verdict came back that this was a completely new family of wildlife unknown to science, whereupon it was given a formal name and treated to media fanfare.[2081]

The same year, an exceptionally well-preserved fossil from a rodent family that allegedly went extinct 11 million years ago was uncovered, and someone had the idea to compare it with the Laotian rock rat. When this was done, they proved to be "a striking match."[2082] [2083] [2084] Slight variations such as the number of roots in their upper molar teeth were found,[2085] but as will be shown, we humans exhibit far more variation than this.[2086]

It's easy to lose perspective when timeframes in the millions of years are spelled out with words instead of digits, so let's rectify this. This family of wildlife was said to have gone extinct 11,000,000 years ago, or a 1,000 times further back in time than when the saber-toothed tiger died out 10,000 years ago.[2087] Interestingly, scientists label such finds with the phrase "Lazarus Effect," referring to the man Jesus raised from the dead in the village of Bethany.[2088] [2089] This happens to be the same term that the scientists who made this discovery used in the title of their paper.[2090] From any rational perspective, this does not bode well for the geologic time scale or the theory of evolution, yet it pales in comparison to what follows.

Long Before Dinosaurs

Going back to at least 1823, fossils of a strange, menacing family of fish began to be unearthed in the rocks of Europe.[2091]

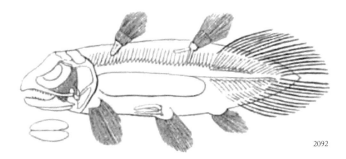

2092

Fossils of this creature, later named the coelacanth (pronounced seel-a-kanth), were discovered across the globe, and a puzzle became immediately apparent. Despite the fact that they were found in various different rock formations dated from 66 to 385 million years ago, all of the fossils were extremely similar.[2093] [2094] How could it be that over such an "inconceivable duration of successive geological ages," as one evolutionist phrased it, this "elaborate and complicated" creature had barely changed?[2095] (1873) Surely there was variation,[2096] but as explained in several academic texts:

- over "space" and "time," there was a "remarkable persistence" of "even minute characters" (1873)[2097];
- the older forms were "as completely developed" as the newer ones (1908)[2098];
- "all the essentials of its very peculiar structure" were "identical" (1862)[2099], and
- the family remained "practically unchanged." (1904)[2100]

Despite this, instead of being viewed as a challenge to the theory of evolution, the coelacanth was simply labeled with the term "persistent type" and shoved into the evolutionary framework. (1866)[2101] So in sum, the evolutionary interpretation of the fossil record was that over a period spanning some 300,000,000 years that began 150,000,000 years before the first dinosaurs emerged, coelacanths remained "practically unchanged" until they disappeared 70,000,000 years ago when "dinosaurs still ruled the earth."[2102] [2103] [2104] [2105] [2106] Hence, the artist's interpretation below:

© iStockphoto.com/AdrianChesterman

If it seems implausible that coelacanths existed and experienced so little change over such an incomprehensible period of time, be prepared for what is next. In the city of East London, South Africa on December 22, 1938, a trawler captain summoned a local museum curator to examine an unusual fish caught that morning. This fish, which the curator described as a "most queer-looking specimen," turned out to be a living coelacanth.[2107] [2108] Being that coelacanths were supposed to have gone extinct in the age of the dinosaurs, the discovery created quite a stir, and some initially refused to believe it.[2109] Since then, however, all doubt has been removed as more than 150 living coelacanths have been pulled from the waters (see picture on next page).[2110] [2111]

Most remarkably, in the words of the scientist who first identified the living coelacanth:[2113]

> The bony structures of our modern Coelacanth are almost exactly the same as those left by Coelacanths of several hundred million years ago.[2114] (1956)

Likewise, renowned palaeontologist Niles Eldredge explained that the

Ronan Bourhis/AFP/Getty Images

Coelacanth caught off the coast of Indonesia in 2007[2112]

coelacanth that lived among the dinosaurs is "in many ways the spitting image" of the coelacanth that lives right now, except it was "less than a foot in length" and thus "only a fraction as long as its four-foot living relative." (1991)[2115] [2116] Yet, even this lone distinction he draws is inaccurate. The scientist who first discovered the dinosaur-era coelacanth reported back in 1851 that it was "between two and three feet in length" at maturity.[2117] [2118]

Again, there is certainly variation between the living coelacanths and fossils, but this is to be expected. All life forms exhibit variation, which, as we saw in our study of genetics, is not to be confused with evolution. Take humans for example. The world's tallest man is 7 feet, 9 inches tall, while the shortest is 2 feet, 5 inches (see picture on next page).

Imagine if the bones of these men were found in different rock formations in different parts of the globe. There can be little doubt they would be labeled as different species. Yet, we know for a fact they are both human and both live at the same time. Furthermore, both men are of the same race, and both were born in Inner Mongolia.[2120] Even though we are all clearly human,[2121] variations among people are often unappreciated. One clinical textbook fills nearly 600 pages with explanations, drawings, and pictures of differences between individuals such as the number of bones in our vertebral columns, the routings of our arteries, and the shapes of our brains. The preface of this book notes that "students are often frustrated because the bodies they are dissecting do not conform to atlas or textbook descriptions." More importantly, it states such descriptions "are accurate or hold in only 50–70% of individuals," and thus, when it comes to practicing medicine, they "are not only inadequate but may be dangerously misleading as well." (1988)[2122]

© 2009 Associated Press

July 13, 2007: The world's tallest man, Bao Xishun, shakes hands with He Pingping, who measures only 2 feet, 5 inches.[2119]

Yes, species vary, and sometimes quite significantly as we see with humans, dogs, and many other forms of life. Yet, fossils found in different rock strata with only "trivial" variations or none at all have been labeled as different species. (1878)[2123] (2003)[2124] (1984)[2125] (1984)[2126] Darwin left us incisive testimony to this fact when he wrote:

> It is notorious on what excessively slight differences many palae-ontologists have founded their species; and they do this the more readily if the specimens come from different sub-stages of the same formation. (1859)[2127]

This is another example of circular logic founded upon evolutionary pre-sumptions. Since it is generally assumed that creatures separated by strata are separated by millions of years, it is also assumed that they must be different species—even if, as with coelacanths, they look "almost exactly the same."[2128]

PRETENDING THE CONFLICT DOES NOT EXIST

In his 1991 book, *Fossils: The Evolution and Extinction of Species*, Niles Eldredge wrote that "rocks are about the same age if they contain the same, or closely similar fossils." (1991)[2129] As we see, however, this does not hold true when it comes to mussel shrimps, grasses, rock rats, coelacanths, and numerous other currently living organisms, just a few of which are outlined below:

- *Ginkgo biloba* trees, which are "indistinguishable" from fossils dated to more than 100,000,000 years ago (2003)[2130]
- Blue coral, which cannot be differentiated from fossils alleged to be 100,000,000 years old (1984)[2131]
- Cryptobranchid salamanders, "whose structures have remained little changed for more than 160,000,000 years" (2003)[2132]
- Lampreys, which have remained "astonishingly stable" for 360,000,000 years (2006)[2133 2134 2135]
- Horseshoe crabs, which are "strikingly similar" to recently discovered 445,000,000 year-old fossils (2008)[2136 2137]
- The shellfish *Lingula*, which has "has remained virtually unchanged for the entire Phanerozoic [542,000,000 years ago to the present]!" (2008)[2138 2139]

Since these organisms are all alive right now, what are we to make of the claim that "rocks are about the same age if they contain the same, or closely similar fossils"? Either this statement is inaccurate, or rocks thought to be hundreds of millions of years old are in fact much younger. Either way, there is a massive disconnect between theory and fact. Furthermore, these examples represent the mere surface of fossil evidences that undercut the theory of evolution. The more indicting and widespread problems are brushed aside in the ways described by renowned physicist Robert Jastrow:

> It turns out that the scientist behaves the way the rest of us do when our beliefs are in conflict with the evidence. We become irritated, we pretend the conflict does not exist, or we paper it over with meaningless phrases. (1978)[2140]

Observe how this plays out. The Encyclopædia Britannica states that mussel shrimps "often" serve "as index fossils owing to their abundance, widespread geographic occurrence, and limited vertical range." (2004)[2141] Recall from above that "whenever a rock is found bearing" an index fossil, "its approximate age is automatically established." (1977)[2142] How is it, then, that the discovery of an exquisitely preserved, 425 million-year-old mussel shrimp that is "almost exactly" like those living today does not throw this doctrine into disrepute?[2143 2144] An article in the journal *Science* coinciding with the publication of this discovery notes geologists use mussel shrimp shells "to date and analyze rocks," but there isn't even a hint that this practice should be critically reexamined in the wake of this finding. (2003)[2145] Think about it. If the oldest mussel shrimp fossil that preserves fine details of its appearance looks "eerily similar" to living mussel shrimps,[2146] [2147] how can mussel shrimps plausibly be used to date rocks?

In his 2001 book, Ernst Mayr, the "Darwin of the 20th century," writes:

> The most convincing evidence for the occurrence of evolution is the discovery of extinct organisms in older geological strata. … The older the strata are in which a fossil is found—that is, the further back in time—the more different the fossil will be from living representatives.[2148]

This statement is in clear disaccord with the 2003 mussel shrimp discovery. So much for the "most convincing evidence" for evolution. The same applies to the 445-million-year-old horseshoe crab fossils detailed in a paper published in 2008. According to evolutionists, these fossils are:

- older than any other horseshoe crab fossils by 95,000,000 years,[2149]
- older than the creatures once alleged to be the ancestors of horseshoe crabs,[2150] [2151] and
- "much closer" in appearance to living horseshoe crabs than to horseshoe crabs fossils dated at 335,000,0000 years ago.[2152] [2153]

How is all of this rationalized? With the statement that the modern horseshoe crab body plan must have "evolved considerably earlier … than was previously suspected."[2154] Talk about pretending a conflict doesn't exist. If inconsistencies of this magnitude can be brushed aside with such rhetoric and without causing evolutionists to question their interpretation of the fossil record, one has to wonder if they even allow themselves the intellectual freedom to do so. Sure, they'll question the details, but to question the big picture seems forbidden.

A similar scenario exists with lampreys, which are alive right now, but "are unknown as fossils" for the past 300,000,000 years. (1998)[2155] Yet, "exceptionally preserved" 330 million-year-old lamprey fossils have been found that are "almost identical" to living lampreys.[2156] If this isn't enough, a recently discovered 360-million-year-old lamprey fossil hails from the same period as the creatures it was supposed to have descended from, is 35,000,000 years older than the next oldest lamprey fossils, and is even more similar to living lampreys than they are. (2006)[2157] [2158]

Again, all this is brushed aside with the claim that lampreys must have evolved earlier than thought.[2159] Sure, and grasses, mussel shrimps, and horseshoe crabs must have evolved earlier than thought too, while coelacanths and rock rats must have gone extinct later than thought—or, for that matter, not at all. All of this goes to show that the evolutionary view of the fossil record embodies the definition of pseudoscience: "Either the contrary data are ignored, or new details are continuously added to the theory in order to explain all new observations."[2160] Worse yet, this is not an "either/or" scenario, for the ignoring of contrary data and the continual addition of new details are both at play.

We're hardly done here because the problems for the theory of evolution run far deeper. The findings we have been discussing have come to light only because previous conclusions about the fossil record painted the evolutionary storylines into corners. For instance, if the 425 million-year-old mussel shrimp fossil was found in a rock formation that was not previously dated, both the rock and fossil would be dated as modern based on its resemblance to living forms. Remember, the fossils typically date the rock, not vice versa.[2161 2162 2163 2164] In which case, the fossil would disappear into a collection or be discarded as mundane.

There are many other such opportunities for theory to lose touch with reality because evolutionists have a habit of invoking hypothetical scenarios to rationalize evidence that doesn't fit with their views. A 1944 work dedicated to the subject of index fossils and written by a professor of palaeontology and a professor of geology from MIT lays the groundwork for understanding how and why this occurs:

> Identifications and correlations should always, of course, be based on a large and well preserved assemblage of characteristic fossils, but that situation is too infrequently encountered in actuality. More often the paleontologist has only a few fairly well preserved specimens, and frequently he may have to determine the identity of a fossil or the age of a formation by means of a few scraps or fragments which have escaped destruction by scavengers, diagenesis, recrystallization, or weathering.[2165]

The sum of this is that formations are frequently correlated on the basis of limited fossil evidence. Yet, some paleontologists openly admit such evidence can be dismissed if it conflicts with theory. For example, in propounding a theory about dinosaurs, David Raup, whom Stephen J. Gould referred to as "the world's most brilliant paleontologist," writes:

Odd fragments and isolated bones are ruled out in this case, because the chances are so good that isolated pieces may be buried, exhumed by erosion, and deposited again long after the animal actually died. (1986)

Hence, Raup confesses his theory:

requires a certain amount of inference, although this kind of inference is standard in field geology. (1986)[2166]

Yes, this kind of inference is standard, and such conjecture is repeatedly used to explain away findings that theory says should not exist. This includes:

- the "remarkable" and "surprising" discovery of fossils in the same rock sample separated in time by more than 100,000,000 years (1933)[2167 2168 2169];

- spores in 250,000,000-year-old salt crystals that grow living bacteria when cultured (2001)[2170 2171 2172 2173 2174];

- two-billion-year-old rocks that appear younger than those that "have survived only a few million years" (1971)[2175];

- a geological formation where clays separated by more than 488,000,000 years sit one on top of another without visible distinction between them (2004)[2176 2177];

- thirty separate fossil zones crammed into the height of one foot in Sicily while just one of these same zones is 15,000 feet thick in Oregon (1973)[2178 2179];

- wood and spores from land plants dating back to the Cambrian period [488–542 mya] while land plants were not supposed to have evolved until the Upper Silurian Period [≈ 416–423 mya] (1974)[2180];

- adjoining layers of limestone in the Grand Canyon that contain fossils separated in time by three full geological periods but show less evidence of separation between them than limestone layers confined to a single geological sub-period (1957)[2181], and

- a rock unit in the Grand Canyon that exhibits "perfect continuity" over its entire length but "is of different ages at the two ends of the canyon, so that physical continuity has failed completely to establish correlation." (1957)[2182]

All the more outrageous, the vast majority of such findings are never published because they are dismissed in the field.[2183 2184] This, of course, only applies to discoveries that conflict with theory. When discoveries are consistent with theory, then, as explained above, "a few scraps or fragments" of fossils or mere inference is sufficient to draw conclusions. This is especially troubling in light of the fact that during the era from 251–542 mya, which constitutes more than half of the Phanerozoic, "most species and genera are reported from only a single locality." (1988)[2185] How illogical is it, then, to set aside evidence that doesn't fit with one's views? To justify this double standard, evolutionists would undoubtedly retort that "surprising" discoveries should be greeted with skepticism precisely because they are anomalous with other evidence. Not so. They are only anomalous with the evolutionary interpretation of other evidence, which is grounded in a 150-year-old paradigm built on patently erroneous assumptions.

All this is not to say there aren't sometimes justifiable explanations for "remarkable" discoveries, but in every one of the examples cited above, not a single one is given. In fact, if you examine the citations, you'll see the only basis for attempting to rationalize these findings is that they don't coincide with evolutionary views of the fossil record. Moreover, you will also see how attempts to explain away the findings are made in the face of admitted evidence to the contrary. In summary of everything revealed thus far about these matters, no words could be more applicable than those of Stephen J. Gould:

> Theory must play a role in guiding observation, and theory will not fall on the basis of data accumulated in its own light.[2186]

This is a clear admonition, and what it calls for is exactly what's missing in the evolutionary mindset: a broad look at the situation. Many intellectuals have a way of missing the forest for the trees. Hence, in a book that lays out evidence for catastrophism most evolutionists have been oblivious to, evolutionary geologist Derek Ager writes:

> It seems to me that the conclusions contained in this book are inescapable, if one is not too involved in the minutiae of stratigraphical correlation actually to see them. (1973)[2187]

Making a similar point in a comical way, MIT physicist Victor Weisskopf is credited with saying, "An expert is someone who knows more and more about less and less, until finally he knows everything about nothing."[2188]

CAUSES OF PATTERNS IN THE FOSSIL RECORD

As William Smith discovered two centuries ago, fossils tend to be vertically sorted in patterns, but since the same erroneous notions debunked
above have been used to interpret these patterns since almost two centuries ago, let's examine them anew with the benefit of more light. First and
foremost, we must come to grips with the reality that geological layers do
not provide records of everything that existed on earth during particular
eras. Take, for example, William Smith's ninth layer, which was characterized by a few shellfish and some wood:

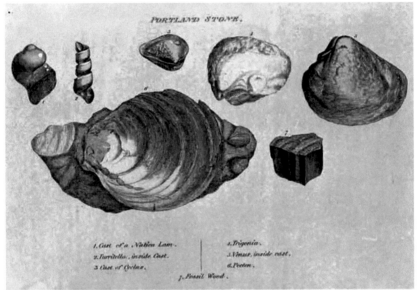

2189

What we see here is obviously not a broad representation of life on
earth during some era of the past as Charles Lyell assumed. (1842)[2190] It
isn't even a record of such for a brief moment in time confined to the vicinity of England. Of the 34 layers into which Smith divided the geology of
England, the maximum number of life forms he identified for any one of
these was 21.[2191] Yet untold thousands of different organisms inhabit England and its surrounding waters at this very instant. Why the discrepancy?
First, as explained by A. B. Smith of the Department of Palaeontology
at the Natural History Museum of London, "Only a small proportion of
organisms ever enter the fossil record." (2001)[2192] Saying much the same a
little more profoundly, Derek Ager writes:

[E]arth history is not a record of what actually happened, *It is a record of what happens to have been preserved…. (1993)*[2193]

Let's get even closer to reality by noting that earth history, or at least as much as anyone can know of it, is a record of what happens to have been preserved, discovered, and published. Moreover, what happens to have been preserved and where it happens to have been preserved are impacted by numerous factors that evolutionists have routinely failed to account for. As M. Foote of the Department of Geophysical Sciences at the University of Chicago admits after drawing some conclusions about diversity in the fossil record:

> These models may seem burdened with unrealistic assumptions, but they are no more so than taking the fossil record at face value, which tacitly assumes a complete, or at least unbiased fossil record." (2001)[2194]

The tacit assumption that the fossil record is unbiased has so permeated evolutionary thought that it is used as a benchmark to justify other unrealistic assumptions. Yet as explained by A. B. Smith, there are a "bewilderingly large number of biases that act upon it." (2001)[2195] For starters, the *McGraw-Hill Encyclopedia of Science & Technology* states only about one-tenth of existing life forms "have tissues and life habitats that make them readily fossilizable." Furthermore, the encyclopedia explains that this small minority of organisms generally consists of those with hard parts such as exoskeletons and bones, and "[u]ncritical reliance on such skeletal fossils leads to grave biases with regard to the nature and anatomy of individual organisms and to the composition of the once-living biota." (2007)[2196]

Besides the possession of hard parts, other significant factors that impact the chances for fossilization and/or the vertical order in which fossils are sorted include:

- physical size (smaller organisms are more likely to be fossilized because they are more quickly buried),[2197]
- general habitat (since the "great majority of all fossils are preserved in water-borne sediments," organisms that live in or near water are more likely to be fossilized),[2198 2199 2200 2201]
- specific habitat (as in the Hall Canyon of California, where aquatic organisms appear in differing strata according to the depths at which they lived),[2202]
- post-mortem buoyancy (creatures that float after they die are apt to

decay before they can fossilize),

- hydrodynamic properties (bones are sorted by water according to their density, size, and shape),[2203] [2204]

- speed and intelligence (faster and smarter land creatures are less likely to be entombed by encroaching floodwaters; more mobile sea creatures are less apt to be buried in underwater landslides),

- population sizes (organisms that are more populous at various points in time are more likely to be fossilized and subsequently found),[2205] [2206]

- migratory capacity (organisms that can easily cross physical barriers such as oceans are apt to be the first to colonize newly formed lands and thus appear lower and more often in the fossil record),[2207] and

- local and global environmental conditions (as these vary, plant and animal populations increase/subside/migrate).[2208] [2209]

It's not that evolutionists are unaware of these factors—it's that they have conflated their sweeping effects with notions about how "life had changed over time." (1999)[2210] To quote more fully A. B. Smith's words cited earlier:

> Although the fossil record provides direct evidence for how bio-diversity has changed over time, it cannot be necessarily taken at face value because of the bewilderingly large number of biases that act upon it. A proper appreciation of those biases is only now beginning to emerge. (2001)[2211]

Only now beginning to emerge? How about all the conclusions drawn over the past 200 years? Maybe this is why the closing paper of a major serial work on paleontology issues a "warning" that "many aspects of the occurrence of fossils cannot be read as simply as previously interpreted." (2001)[2212] Remember, living at this very moment are creatures such as lampreys, that don't show up in the fossil record until three hundred million years ago.[2213] [2214] Think about it. If lampreys had gone extinct a few thousand years ago, evolutionists would be pointing to the fossil record and pompously lecturing us that lampreys could not have possibly lived at the same time as humans—or mammals—or birds—or flowering plants —or dinosaurs—because none of these life forms existed until long after lampreys disappeared from the face of the earth.[2215] [2216] Meanwhile, when "greatly surprised" evolutionists find in the same rock sample creatures that are supposedly separated by millions of years, they are dismissed in the field by the "average worker" who is "familiar" enough with the fossil record to know that they could not have possibly coexisted.[2217] [2218] (1933)[2219] Put simply, evolutionists are arguing from silence[2220] [2221] when evidence and logic show it makes little sense to do so. Not only that, but in certain known cases and an inscrutable number of others, the silence is of their own making.

Reconstruction and Interpretation

Regrettably, evolutionary warping of the fossil record is not restricted to the "where," "why," and "when" of fossils but actually extends to the "what" as well. Consider cases such as these:

- Alleged remains of living cells in the "world's oldest rocks" were found to be crystals "rusted by water that has seeped into the rock." (1981)[2222]

- "Obese volumes" of scientific literature were written about a fossil tooth that showed "Darwin was right" and proved the existence of a primitive "ape-man." (1922)[2223] (1922)[2224] (1925)[2225] Five years after this fossil was unveiled to the world, it was "positively identified" as a tooth from an "extinct wild pig." (1928)[2226]

- Conclusions are still drawn from tenuous reconstructions and small fossil fragments such as teeth (1988)[2227] (2001)[2228] (1999)[2229] (2004)[2230], even though it has been repeatedly shown such extrapolations are manifestly fallible. (2005)[2231] (2005)[2232] (1991)[2233] (1985)[2234] (2002)[2235]

- The discovery of an "apelike and speechless" man buried with "flint and bone tools" was trumpeted by the *New York Times* with the headline, "Darwin Theory is Proved True." The find was called "the missing link in the chain in man's evolution" and the discovery that "takes us back nearer to the source and origin of the first living creature than any other discovery ever made." (1912)[2236] (1912)[2237] (1912)[2238] (1913)[2239] [2240] Forty years later, this entire set of "famous" fossils was exposed as a "hoax" incorporating the jawbone "of a modern ape, probably an orangutan, that has been 'doctored' with chemicals to give it an aged appearance." (1953)[2241] (1953)[2242] (1953)[2243] (1954)[2244]

- In 1978, a family of Jurassic birds was established based upon a single fossil alleged to be "intermediate between a reptilian scale and a feather." Since then, at least four scientists have declared it is nothing more than the leaf of a plant. (1999)[2245] (2004)[2246]

Perhaps worse than any of the above was widespread use of the absurd myth that "geological succession and modern embryological succession have near parallelisms." To quote from an 1895 geology text, "Paleontologists of skill derive a degree of prophetic power through the aid of the canon."[2247] As elucidated by P. J. Bowler of the Department of Social Anthropology at Queen's University:

> Haeckel's recapitulation theory—the claim that ontogeny recapitulates phylogeny—was widely accepted by palaeontologists looking for clues as to the 'shape' of the pattern they should expect to find. (1990)[2248]

In what can hardly be coincidence, this practice reflected a hope Darwin voiced in the conclusion of the *Origin of Species*: "If it should hereafter be proved that ancient animals resemble to a certain extent the embryos of more recent animals of the same class, the fact will be intelligible." (1859)[2249] Thus, "between 1860 and 1920," as explained in the book, *Historical Geology*, this "idea of recapitulation affected every branch of paleontology...." (1965)[2250]

Far worse, in modern evolutionary texts written by Grassé, Gould, and Eldredge, variations of this fairy tale are cited to fill holes in theories that they explicitly confess are unsupported by the fossil record.(1973)[2251] (1977)[2252] (1977)[2253] (1991)[2254]

There is no way of knowing how many other cases exist wherein fossil evidence is mangled by evolutionary presumptions, but given the simplistic and obvious nature of these examples, there are undoubtedly manifold others. This is not about conspiracy. Nor is it about random cases of dishonesty or ineptitude. It is about masses of intelligent and educated people being so eager to believe in evolution that appropriate scientific caution is cast aside. As we've seen time and time again, evolutionary interpretations of the fossil record are so pliable that expectations have a way of becoming self-fulfilling prophecies, even when totally removed from reality. In the words of Jeffrey H. Schwartz, Ph.D. in anthropology from Columbia University and professor of anthropology at the University of Pittsburgh:

> If you impose your view of the process of evolution on the fossils, then, of course, the picture you will get will conform to your expectations. (1999)[2255]

Such vast opportunity for such prejudice to take hold is made worse when fossils are fragmented or poorly perserved, leaving plentiful room to the imagination. In contrast, the striking correspondences between living organisms and some of the fossils mentioned above were recognized because the fossils were well-preserved or advanced technologies were used to analyze them. This applies to the 425 million-year-old mussel shrimp,[2256] [2257] [2258] 11 million-year-old rock rat,[2259] 360 million-year-old lamprey,[2260] [2261] [2262] and dinosaur feces containing remains consistent with that of living grasses.[2263] [2264] [2265] [2266] In fact, were it not for recent advances in the study of plant remains, evolutionists would still be ranting about the absence of modern foliage in dinosaur dung.[2267] It is noteworthy that such detailed and well-substantiated evidence points to creation, while, on the contrary, the case for evolution is built upon murky data and shaky inferences.

DATING

Throughout this book, the conflicting goals of substantively addressing wide-ranging subjects and keeping this work to a reasonable length have forced me to make difficult decisions about what to cover, but none more so than regarding the topic of dating. We could easily spend several hundred pages on this but instead will cut to the core of it in the space of just a few. Interestingly enough, this parallels the fact that many natural processes presumed to require millions of years can proceed far more quickly than previously thought. As we have already seen, sedimentary rock layers can form much more quickly than previously assumed, and similarly:

- A 2004 paper in the journal *Sedimentary Geology* notes some "researchers believe that several millions of years are necessary for the complete formation" of petrified wood, but experimentation in a natural mineral spring has shown it "can form under suitable conditions in time periods as short as tens to hundreds of years…."[2268]

- As explained in a press release from the University of Toronto, an associate professor of geology at this institution coauthored a paper in the journal *Nature,* finding that:

> the way that granite forms — a rock that makes up about 70 to 80 per cent of the Earth's continental crust—is not the sluggish, multi-million year process that scientists previously believed. In fact, [the professor] and his co-authors argue that the process occurs in rapid, dynamic and possibly catastrophic events that take between 1,000

and 100,000 years, depending on the size of the granite intrusion. (2000)[2269] [2270]

- A 2001 geography text published by Oxford University Press claims stalactites and stalagmites grow at the miniscule rate of 0.04 inches per hundred years.[2271] By this reckoning, it would take 1,500,000 years for a fifty-foot stalactite to form. In reality, however, scientists have known for more than a century that the rates at which stalactites form vary greatly depending on conditions such as rainfall. Stalactite growth rates of five inches/hundred years have been measured in caverns along the east coast of the United States and 30 inches/hundred years in caves located in Great Britain, at which rate a 50 foot stalactite will form in 2,000 years. (1889)[2272] (2007)[2273] (1898)[2274]

- A 2004 article in the magazine *New Scientist* reports that a "fully functioning cloud rainforest ecosystem" has developed on the mid-Atlantic island of Ascension "virtually from scratch in just 150 years." Darwin himself visited this island in 1836 and wrote that it was "entirely destitute of trees." Now scientists are contending with the fact that:

> according to ecological theory, rainforests are supposed to evolve slowly over millions of years, as species co-evolve and ecological niches are created and filled. Discovering the Green Mountain cloud forest is like finding that a pile of used car parts in a scrapyard has spontaneously reassembled into a functioning car. Unless, that is, ecologists have gotten their theories hopelessly wrong. (2004)[2275]

- Through a series of volcanic eruptions, an island called Surtsey was born off the coast of Iceland in the 1960's. Now, fewer than 50 years later, it is home to a "fully functioning ecosystem." As explained in *New Scientist*: "All" have been "confounded" by what has taken place, and scientists are stunned by the "speed, ingenuity and sheer unpredictability of nature's colonization" of the island, all without any "complex evolutionary adaptation to the surroundings...." Even more significantly:

> The island has excited geographers, who marvel that canyons, gullies and other land features that typically take tens of thousands or millions of years to form were created

in less than a decade. (2006)[2276]

This is only a mere sampling of findings that illustrate how common knowledge about processes requiring millions of years relies on uncritical assumptions that have not been borne out by the sole judge of scientific truth. With this, let's examine the methodology of dating fossils.

The most well-known of all dating techniques is carbon dating, otherwise known as radiocarbon or ^{14}C dating.[2277] The sure mark of an amateur evolutionist is to assert that carbon dating shows certain fossils are millions of years old. Nothing could be further from the truth. Carbon dating is based upon the decay of carbon-14, a radioactive element that is present in the air and ingested by living organisms. After an organism's death, carbon-14 intake generally stops, while that which is present in the organism decays. Hence, by measuring the amount of carbon-14 in a fossil, a date of death can be calculated.[2278] The problem for anyone who thinks this process can show a fossil to be millions of years old is that carbon-14 has a half-life of 5730 ± 40 years.[2279] Hence, in twenty such periods (about 115,000 years), a given amount of carbon-14 is halved twenty times, leaving less than 0.0001% of it,[2280] which is not enough to perform a dating measurement. As Simon & Shuster's *New Millennium Encyclopedia* explains:

> The rapid disintegration of carbon-14 generally limits the dating period to approximately 50,000 years, although the method is sometimes extended to 70,000 years. Uncertainty in measurement increases with the age of the sample. (1999)[2281]

Ignorance about this among the general public is at least partly attributable to prominent media outlets and public figures who have asserted that carbon dating shows objects to be millions and even billions of years old.[2282] [2283] [2284] [2285]

Now here comes a bombshell. Not only is carbon dating unable to show that fossils are millions of years old, it also actively threatens this belief. This is because fossil materials that "should exhibit no ^{14}C activity" due to their "great geologic age" typically do. For nearly thirty years, scientists have been working to identify and eliminate any factors that could cause these unanticipated results. They have even carbon dated diamonds, which have properties that make them ideal for such purposes. The result was that the diamonds yielded carbon-14 dates in the range of 65,000 – 80,000 years old, despite the fact that they are alleged to be hundreds of millions of years old on the basis of their geological context. Such results leave only two possibilities: Either these organic materials are far younger than assumed, or extenuating factors have skewed the carbon dating tests. Being unwilling to accept the former, evolutionists assume the latter even though they have taken extraordinary measures to eliminate any extenuating factors that could skew the tests. (2007)[2286] (2007 – creationist source)[2287]

Moreover, evolutionists employ such logic on an inconsistent basis. Creationists have conducted blind tests in which they send fossil materials to highly regarded independent labs that perform precision carbon dating. They have used different labs and different types of fossil materials, but the results are consistent: Materials that are supposed to be millions of years old based on their geological context are carbon dated to a tiny fraction of this timeframe.[2288] [2289] [2290] [2291] The inconsistency is that when the creationists ask the labs about the possibility of contamination and other factors that could distort the results, the labs assure them of the accuracy of the dates.[2292] [2293] However, when the results are published and the conflict between the radio-carbon dates and geological ages are revealed, evolutionists revert to special pleadings in attempts to explain away the results.[2294] The double standard is obvious, but such is the power of entrenched paradigms.

For dating fossils theoretically older than carbon dating allows, there are numerous radiometric dating methods based upon elements with half-lives much longer than carbon-14. These elements, however, are typically found in volcanic materials, which seldom contain fossils because the intense heat of volcanic activity destroys them. (2007)[2295] (1993)[2296] Hence, the process of dating such fossils is typically one of inference in which the fossils are not directly dated. Rather, volcanic rocks or ash that have been correlated to fossils are dated. If you're wondering how much inference is involved in such correlation, a college textbook explains that:

> datable ash layers rarely fall out of the sky because stratigraphers need them! More commonly, rocks for which absolute ages can be obtained are widely separated, not only in time, but also geo-graphically. (1996)[2297]

Contrary to popular perception, scientists generally do not place a fos-sil in a machine and get a date out. More commonly, they place several volcanic rock samples that have been correlated to a fossil in a machine, get a variety of dates out and then interpret the evidence. As explained in a book about the geologic time scale published by Cambridge University

Press, "[I]t is well not to refer to radiometric ages as absolute ages" because they are not actually "*true*" ages, but "*apparent*" ages. (1982)[2298] This reality becomes even more vivid in light of these academic texts:

> Often it is not easy to reconcile different [radiometric] dates, and many age determinations which do not agree with currently accepted time scales are simply rejected as wrong without our fully understanding why.
> – *The Natural History of Fossils* (1980)[2299]

> Absolute dating methods such as radiocarbon analysis are of much utility in floodplain stratigraphic work. However, such analyses are expensive, and the needed sample materials are not present in every exposure. Moreover, absolute dating methods are all subject to various possible errors, making it desirable for stratigraphic mapping to be conducted first so that samples submitted for dating can be placed in their correct, relative age, order.
> – *Flood Geomorphology* (1988)[2300]

> [D]ifferences (often essential) between data determining absolute age and geological data is a usual phenomenon that surprises nobody; it has always been solved in favor of the latter. In the absence of geological data, determination of absolute age often leads us into great errors.
> – *Critical Aspects of the Plate Tectonics Theory* (1990)[2301]

> Geologists put more faith in the principles of superposition and faunal succession [the geological column] than they do in numbers that come out of a machine. If the laboratory results contradict the field evidence, the geologist assumes that there is something wrong with the machine date. To put it another way, "good" dates are those that agree with the field data.
> – *Cascadia: The Geologic Evolution of the Pacific Northwest* (1972)[2302]

> It is obvious that radiometric techniques may not be the absolute dating methods that they are claimed to be. Age estimates on a given geological stratum by different radiometric methods are often quite different (sometimes by hundreds of millions of years). There is no absolutely reliable long-term radiological "clock."
> – *The Science of Evolution* (1977)[2303]

"Different" by "hundreds of millions of years"? For perspective, humans and kangaroos allegedly share a common ancestor that lived 135 million years ago.[2304] Given such disparities in radiometric ages, perhaps this evolutionary divergence only occurred a few generations ago and that is why I have a large snout and sometimes feel an insatiable urge to box. In total, the academic sources above make it glaringly clear that despite habitual usage of the word "absolute" in reference to radiometric dating techniques, this term is a misnomer.

Another potentially misleading element of radiometric dating parlance is the plus or minus symbol (±) that commonly accompanies such dates. As explained by Neil de Grasse Tyson, Ph.D. in Astrophysics from Columbia University and research scientist at Princeton University: "The entire (responsible) scientific community, when reporting a measurement or computation, will give the reader an indication of the believability of the results." Towards this end, he suggests using the ± symbol, which is "more than a math symbol," but "a symbol of honest uncertainty." (1993)[2305]

Lamentably, this is not the case when it comes to radiometric dating. As explained by Steven M. Stanley of Johns Hopkins University, the plus or minus figures in radiometric dates do not account for potential errors that "sometimes add up to sizable total errors, especially when very old rocks are being dated." (1993)[2306] In the same context, the aforementioned Cambridge University Press book on the geologic time scale avows: "The expression ±x is misleading if given without qualification." (1983)[2307] Though scientists familiar with radiometric dating techniques understand this, the average person does not, and when he sees the mathematical symbol for "honest uncertainty" followed by a number, he assumes all uncertainty has been accounted for when in fact it has not.

The source of this uncertainty and the Achilles' heel of all radiometric dating methods boils down to one word: assumptions. The ± symbol does not represent the accuracy of a date but the accuracy of a dating method if all of the assumptions underlying it hold true.[2308] For example, William D. Stansfield, Professor of Biological Sciences at California Polytechnic State University, outlines the assumptions behind the uranium-238 dating technique, which is based upon the decay of uranium (U^{238}) into lead (Pb^{206}):

If we assume that (1) a rock contained no [lead] when it was formed, (2) all [lead] now in the rock was produced by radioactive decay of [uranium], (3) the rate of decay has been constant, (4) there has been no differential leaching by water of either element, and (5) no [uranium] has been transported into the rock from another source, then we might expect our estimate of age to be fairly accurate. Each assumption is a potential variable, the magnitude of which can seldom be ascertained. (1977)[2309]

Take special note of that last sentence because it concisely summarizes the underlying pitfall of radiometric dating. Yet, it has been claimed that certain dating methods allow us to firmly constrain such variables. Dr. Ken Miller, for instance, writes:

The potassium-argon method is popular with geologists because the only serious source of error arises when a rock has been disturbed by heating, allowing the accumulated argon gas to escape. In all such cases, the time span determined is an underestimate of the rock's true age, never an overestimate." (1999)[2310]

Such cocksure yet unsubstantiated assertions are invariably deflated by reputable sources. Consider the following from a text on radiometric dating published by Oxford University Press:

Almost immediately upon the development of the [potassium-argon] dating method came reports documenting the presence of excess [argon] in minerals…. In some cases, [potassium-argon] ages much greater than are geologically plausible are observed, and in a few cases they are older than the age of the Earth. (1999)[2311]

The contrast between Dr. Miller's claim and reality couldn't be more stark, and this is not an isolated case. In the same book, Dr. Miller gives an even stronger vote of confidence to a dating technique known as rubidium-strontium isochron analysis,[2312] but, as above, his contention is directly refuted by reputable sources. (1990)[2313] (2005)[2314] Once more, let me emphasize none of this is meant to single out Dr. Miller because comparable examples could be drawn from the writings of many academics. Indeed, such glaring errors are an almost unavoidable consequence of the evolutionary/long-age dogmas that pervade and taint otherwise-credible scientific literature.

Of the five assumptions listed above that impact the reliability of radiometric dating, the most critical of these is that "the rate of decay has been constant." As Stansfield explains:

The uncertainties inherent in radiometric dating are disturbing to geologists and evolutionists, but their overall interpretation supports the concept of a long history of geological evolution. (1977)[2315]

This "overall interpretation" is entirely dependent on the assumption that the decay rates of radioactive materials have not changed over the course of earth's history. It is generally accepted among evolutionists that this is so, and here is a typical explanation of why from a scientist with a Ph.D. in geochemistry from MIT:

> Because no scientist has ever detected a significant variation in the rate of radioactive decay, no matter the heat or the pressure to which they subjected the parent isotope, we can consider [this] assumption validated. (2005)[2316]

It is correct that heat and pressure have never been shown to significantly change the rate of radioactive decay,[2317] [2318] but why should this question be limited only to heat and pressure? The fact is that decay rates can and do change with exposure to subatomic particles such as those generated by the sun and by the types of chain reactions that power nuclear reactors and atomic bombs.[2319] As explained by Russian scientist N. S. Boganik in a book published by the USSR Academy of Sciences that NASA found worthy of translating and publishing:

> [F]rom the moment of the discovery of artificial radioactivity — and, to an even greater degree, the subsequent breakthrough in the field of nuclear physics culminating in the understanding of chain reactions — any assertion that atomic decay is a purely individual event independent of external influences is simply not in agreement with established facts. ...
> The investigation of cosmic rays has led to direct proof that under natural conditions there are factors which affect the course of natural radioactive transformations. (1970)[2320]

Various scientists such as Boganik (1970),[2321] Einstein (1946),[2322] and Haldane (1944)[2323] have proposed that radioactive decay rates have changed over the course of earth's history. This possibility, however, doesn't appear close to being proved, and I don't pretend to know if such events took place because it is beyond my acumen to develop and constrain the scientific models that could determine this. Nevertheless, I do think it foolhardy to rule out the prospect that decay rates have changed over the course of earth's history given all that is detailed above and the amazing evidence we will now examine.

LONG GONE?

In a 1996 Cambridge University Press textbook about dinosaurs, the coauthors write:

> Dinosaurs last romped on this earth 65 million years ago. This means that their soft tissues—muscles, blood vessels, organs, skin, fatty layers, and so forth—are long gone.[2324]

How then, does one explain these remains discovered inside the thigh bone of a Tyrannosaurus rex?

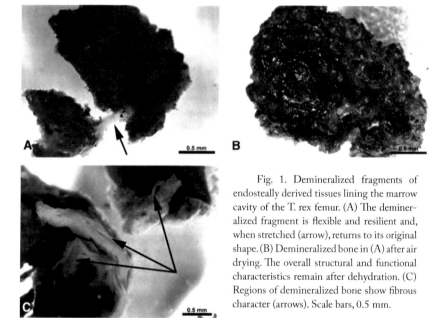

Fig. 1. Demineralized fragments of endosteally derived tissues lining the marrow cavity of the T. rex femur. (A) The demineralized fragment is flexible and resilient and, when stretched (arrow), returns to its original shape. (B) Demineralized bone in (A) after air drying. The overall structural and functional characteristics remain after dehydration. (C) Regions of demineralized bone show fibrous character (arrows). Scale bars, 0.5 mm.

From "Soft-Tissue Vessels and Cellular Preservation in Tyrannosaurus rex." By Mary H. Schweitzer, Jennifer L. Wittmeyer, John R. Horner, and Jan K. Toporski. *Science* **307** (5717), March 25, 2005. [DOI: 10.1126/science.1108397] Reprinted with permission from AAAS.

[2325]

What we see here was revealed after a T. rex femur was accidentally broken and an acidic solution was used to dissolve the minerals in the bone so the interior could be studied more closely. This procedure is normally used to examine soft tissues in modern bones but was rarely utilized on dinosaur bones because it was assumed all soft tissues were long gone.[2326] [2327]

After numerous microscopic examinations and tests, it was found that these bones contain "flexible, hollow blood vessels" that are "similar in all respects to blood vessels" from modern ostrich bones. These similarities include the presence of small, red spheres that look like red blood cells but are "diplomatically" referred to as "round microstructures" out of "an abundance of scientific caution." (2005)[2328] (2005)[2329] (2006)[2330]

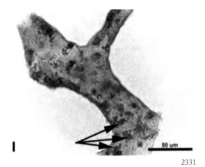

Fig. 2. (I) T. *rex* vessel fragment showing detail of branching pattern and structures morphologically consistent with endothelial cell nuclei (arrows) in vessel wall.

From "Soft-Tissue Vessels and Cellular Preservation in *Tyrannosaurus rex*." By Mary H. Schweitzer, Jennifer L. Wittmeyer, John R. Horner, and Jan K. Toporski. *Science* **307** (5717), March 25, 2005. [DOI: 10.1126/science.1108397] Reprinted with permission from AAAS.

[2331]

Furthermore, although it was thought "impossible" that bones of such age could contain actual collagens (the most common proteins in bones[2332] [2333]), "a variety of chemical and molecular tests" "conducted independently in at least three different labs and by numerous investigators" "strongly support" the conclusion that these bones have "collagen-like protein molecules." (2006)[2334] (2007)[2335] (2007)[2336] (2007)[2337] (2007)[2338] (Also, see refutations of a challenge to these findings and the evolutionary storyline typically associated with them in these citations.[2339] [2340])

That was a mouthful for those who want specifics, but here is the abbreviated version. There shouldn't be any collagen in 65,000,000-year-old bones, but all evidence seems to indicate it is present. The scientists who unveiled this discovery rationalize it by noting that minerals and organic compounds that are intrinsic or find their way into buried bones may enhance the preservation potential of soft tissues such as collagen (2007)[2341], but a paper published in the journal *Paleobiology* noting such phenomena states:

> Even under favorable conditions (cool or dry deposits), however, little intact collagen remains in bones after only [10–30,000 years]. (2000)[2342]

Another paper mentions yet unpublished experimental findings indicating that even if collagen was consistently kept at 0°C, it should be undetectable after 3 million years—a 60 million year disparity between the experimental evidence and evolutionary storyline. Furthermore, at room temperature this detection limit drops to a mere 15,000 years. (2002)[2343]

At even far greater odds with the long evolutionary ages assigned to such fossils is the presence of what appear to be "fragments of hemoglobin molecules." In 1991, Mary Schweitzer, a paleontologist who has been on the leading edge of the discoveries we have been discussing, examined a slice of T. rex bone and found structures that appeared to be red blood cells. In her own words:

> It was exactly like looking at a slice of modern bone. But of course, I couldn't believe it. I said to the lab technician: "The bones, after all, are 65 million years old. How could blood survive that long?" (1993)[2344] (2006)[2345]

Good question. Blood is the most decay-prone tissue in the human body (1974)[2346] [2347], and as explained by Dr. Schweitzer:

> Everyone knows how soft tissues degrade. If you take a blood sample and you stick it on a shelf, you have nothing recognizable in about a week. So why would there be anything left in dinosaurs? (2007)[2348]

Yet while studying under famed paleontologist Jack Horner and earning a Ph.D. in biology from Montana State University,[2349] Schweitzer "analyzed the putative cells using a half-dozen techniques involving chemical analysis and immunology." The results, as summarized in a *Discover* magazine article, were that:

> [a]ll the data supported the conclusion that the T. rex fossil contained fragments of hemoglobin molecules. (2007)[2350] (1997)[2351]

How could such organic materials have survived for 65,000,000 years when we can study their decay in laboratories and find nothing that indicates this is even remotely possible? How implausible do evolutionary paradigms need to become before adherents will question them? As Dr. Schweitzer recounted when she submitted a paper for publication about such preservation to a peer-reviewed journal:

> I had one reviewer tell me that he didn't care what the data said, he knew that what I was finding wasn't possible. I wrote back and said, "Well, what data would convince you?" And he said, "None." (2007)[2352]

On top of this, let's not forget what was revealed earlier in this chapter: coal beds tens of millions of years old with wood "so unaltered that it can easily be cut with a saw and planed"(1953)[2353]; a 40,000,000 year-old fossil deposit containing insects with their original iridescent color and leaves that are still green (1993)[2354]; a 150,000,000 year-old fossil deposit with sea creatures containing ink so unaltered that it can be used for writing when dissolved in water (1990)[2355]; amino acids in fossils said to be more than 440,000,000 years old (2001)[2356] (1990)[2357], and bacterial spores in 250,000,000 year-old salt crystals that grow living bacteria when cultured. (2001)[2358 2359 2360 2361 2362]

All of this points to the obvious conclusion that such fossils may not be as old as imagined. An objective person can hardly avoid it.

"The Trade Secret of Paleontology"

At the outset of this chapter, I made passing mention of what Darwin considered the "most obvious and forcible" objection to his theory. It is now time to see exactly what he was talking about. In the *Origin of Species*, Darwin aptly notes that if his theory is correct, there "must" have been a "truly enormous" number of "intermediate links" tying all life forms to one another. Yet he concedes:

> Why then is not every geological formation and every stratum full of such intermediate links? Geology assuredly does not reveal any such finely graduated organic chain; and this, perhaps, is the most obvious and gravest objection which can be urged against my theory. (1859)

Later in the same book, he reiterates:

> Why does not every collection of fossil remains afford plain evidence of the gradation and mutation of the forms of life? We meet with no such evidence, and this is the most obvious and forcible of the many objections which may be urged against my theory. (1859)

Yet, Darwin didn't consider these facts fatal to his beliefs and put forward the following rationale to explain why:

> I can answer these questions and grave objections only on the supposition that the geological record is far more imperfect than most geologists believe. (1859)[2363]

Darwin was right in one respect. The geological record is extremely imperfect, but as we will see, why is it that with the exception of a few doubtful examples, it is the "intermediate" forms that happen to be missing? Some evolutionists will claim this is not the case, but others, like

Darwin, have been rather candid in admitting it cannot be denied, and some have gone even further by stating Darwin's rationale is not plausible. For instance, one hundred years after the *Origin of Species* was published, the college textbook *General Palaeontology* acknowledged:

> One of the most surprising negative results of palaeontological research in the last century is that such transitional forms seem to be inordinately scarce. In Darwin's time this could perhaps be ascribed to the incompleteness of the palaeontological record and to lack of knowledge, but with the enormous number of fossil species which have been discovered since then, other causes must be found for the almost complete absence of transitional forms. (1959)[2364]

Likewise, Stephen J. Gould, Harvard professor, Ph.D. in paleontology from Columbia University, and "one of the most influential evolutionary biologists of the 20th century," declares:[2365]

> The extreme rarity of transitional forms in the fossil record persists as the trade secret of paleontology. The evolutionary trees that adorn our textbooks have data only at the tips and nodes of their braches; the rest is inference, however reasonable, not the evidence of fossils. …
>
> Darwin's argument [that "the geological record is extremely imperfect"] still persists as the favored escape of most paleontologists from the embarrassment of a record that seems to show so little evidence of evolution. (1977)[2366]

Furthermore, Niles Eldredge, curator of the Department of Invertebrates at the American Museum of Natural History and "one of the leading palaeontologists of our day," writes:

> We cannot blame the lack of transitional forms on a faulty fossil record. (1991)[2367]

Similarly, David B. Kitts of the School of Geology and Geophysics in the Department of the History of Science at the University of Oklahoma explains:

> Despite the bright promise that paleontology provides a means of "seeing" evolution, it has presented some nasty difficulties for evolutionists the most notorious of which is the presence of "gaps" in the fossil record. Evolution requires intermediate forms between

species and paleontology does not provide them. (1974)[2368]

Concurring with all of the above, Collin Paterson of the Paleontology Department of the London Natural History Museum states:

> It seemed obvious to [Darwin] that, if his theory of evolution is correct, fossils ought to provide incontrovertible proof of it because each geological stratum should contain links between the species of earlier and later strata and, if sufficient fossils were collected, it would be possible to arrange them in ancestor-descendant sequences and so build up a precise picture of the course of evolution. This was not so in Darwin's time and today, after many more decades of assiduous fossil collecting, the picture still has extensive gaps. (1999)[2369]

Even more revealing is the nature of these gaps because they are greatest where links are truly needed to support the theory of evolution. The issue is not, for example, whether different types of bears are related to one another. To quote a leading creationist publication, "[A]ll of today's bears probably descended from a single bear kind."[2370] The issue is whether bears are related to ants, and the problem for the theory of evolution is that gaps in the fossil record become even "more universal and more intense" at these levels. Take for instance the polar bear, which is classified as such:

— Kingdom: Animalia
  (as opposed to plants, fungi and types of single-celled organisms)
 — Phylum: Chordata
  (animals that have a nerve cord along their back)
  — Subphylum: Vertebrata (vertebrates)
   — Class: Mammalia (mammals)
    — Subclass: Theria
     (mammals that give birth without an egg)
     — Infraclass: Eutheria
      (mammals that are nourished in utero via a placenta)
      — Order: Carnivora
       (flesh eating mammals such as cats, dogs, and raccoons)
       — Suborder: Caniformia (dog-like carnivores)
        — Family: Ursidae (bears)
         — Genus (plural, genera): *Ursus* (common bears)
          — Species: *Ursus maritimus* (polar bear)[2371 2372]

Observe from above that all bears are part of the same "Family." Then carefully read these words of George Gaylord Simpson—especially the last three sentences:

> It is a feature of the known fossil record that most taxa [all the classifications above such as kingdoms, families, species, etc.] appear abruptly. They are not, as a rule, led up to by a sequence of almost imperceptibly changing forerunners such as Darwin believed should be usual in evolution. A great many sequences of two or a few temporally intergrading species are known, but even at this level most species appear without known *immediate* ancestors, and really long, perfectly complete sequences of numerous species are exceedingly rare. … [T]he appearance of a new genus in the record is usually more abrupt than the appearance of a new species: the gaps involved are generally larger.… This phenomenon becomes more universal and more intense as the hierarchy of categories is ascended. Gaps among known species are sporadic and often small. Gaps among known orders, classes, and phyla are systematic and almost always large. (1960)[2373]

Likewise, Robert L. Carroll, Ph.D. from Harvard, professor of biology at McGill University, and former president of the Society of Vertebrate Paleontology, explains it is fairly easy to imagine all members of the cat family descending from a common ancestor, but when it comes to "evolution between groups with significantly differing ways of life," this "seems to require quite different mechanisms." Thus he draws a "distinction … between these processes, which is emphasized by the terms microevolution and macroevolution." A page later, he revealingly declares that "macroevolutionary events" are "frequently associated with a significant gap in the fossil record." (1988)[2374] This amplifies what he writes at the outset of the same book:

> Where information regarding transitional forms is most eagerly sought, it is least likely to be available. We have no intermediate fossils between rhipidistian fish and early amphibians or between primitive insectivores and bats; only a single species, *Archaeopteryx lithographica* {we'll discuss this creature in a moment} represents the transition between birds and dinosaurs. On the other hand, certain genera of fish, amphibians, and reptiles are known from thousands upon thousands of fossils from every continent. (1988)[2375]

Make no mistake. This is the type of pattern we would expect to find in the fossil record if there is literal truth in the Biblical statement that God made all creatures "according to their kinds."[2376] The Bible does not explicitly define the word "kinds," but based upon the manner in which creatures in the Bible are referred to in common terms such as bears and horses, "kinds" may correspond to somewhere around the level of family in our modern classification scheme. Irrespective of the exact definition, the Bible clearly implies there are boundaries between living creatures, and, moreover, anticipates that the fossil record would not show all creatures smearing into one another as evolutionists imagined. Like Simpson and Carroll, this is precisely what Gould and Eldredge have acknowledged:

> At the higher level of evolutionary transition between basic morphological designs, gradualism has always been in trouble, though it remains the "official" position of most Western evolutionists. Smooth intermediates between [body plans] are almost impossible to construct, even in thought experiments; there is certainly no evidence for them in the fossil record (curious mosaics like *Archaeopteryx* do not count). (1977)[2377]

The common excuse for the absence of such intermediates is that evolutionary transitions occur rapidly and in small populations. Hence, they have very little chance of being preserved and discovered. This hypothesis has been dubbed "punctuated equilibrium," but let's dispense with the academic jargon and evaluate it in plain language because it's really quite simple. This theory asserts that evolution occurs in spurts and through inbreeding. Evolutionists generally avoid the word "inbreeding" when talking about this theory, but when they use phrases like "small, isolated populations", that is exactly what they are talking about. (1977)[2378] (1997)[2379] (1991)[2380] (1982)[2381] (1988)[2382] (2001)[2383] So why not just come out and say it? I'll hazard a guess and say it is because it is generally known that inbreeding has a weakening effect on sexual organisms. Quoting two editions of the college textbook, *Principles of Genetics*:

> Detailed studies have shown that traits such as vigor and fertility decline more or less linearly with the intensity of inbreeding. The more inbred a population is, the less vigorous and fertile it is. (1997)[2384]

> [Breeding in small populations causes a] steady erosion of genetic variability. (2006)[2385]

That's right. To account for a lack of intermediate links in the fossil record, evolutionists are invoking a process that leads to frailty, infertility, and deterioration of genetic diversity. Perhaps evolution really isn't about survival of the fittest but the survival of an unfit theory in the face of evidence to the contrary.

Dubious Intermediates

Despite admitting to the "almost complete absence of transitional forms" in the fossil record, evolutionists have no real choice but to claim they must have existed and will cite a handful of instances as evidence of such. As exemplified in these words of Niles Eldredge:

> Rapid change by no means presupposes that natural selection does not produce an array of intermediate forms between the ancestral and descendant conditions. But rapid evolution, which is most likely to occur in small and rather restricted populations, will tend to reduce the chances of finding traces of the intermediates in the fossil record. …
>
> But intermediates there were, and sometimes we are lucky enough to find them. When we do, they amply validate a basically Darwinian interpretation of large-scale evolutionary change. (1991)[2386]

As a case in point, Eldredge invokes the common claim that mammals evolved from reptiles, and in accord with this, the fossil record shows bones in the jaws of reptiles were transformed into the middle ear bones of mammals. Frequently accompanying this storyline are sketches purporting to depict the transition.[2387]

Having seen several variants of such diagrams, I can't help but be unimpressed. They seem to be a far cry from proving what is alleged. I wondered, however, if I was just being too skeptical until I came across the following statement from Eldredge, and the situation became quite clear:

> The fossil record does not record the details of the final transformation of jaw joint to middle ear bones—but the transition is still carried out in the embryological development of each and every living mammal! (1991)[2388]

In other words, the fossil evidence does not substantiate the claim, but based upon Haeckel's discredited law of recapitulation, it must have happened. Similar irrationalities are associated with other so-called intermediates. *Archaeopteryx*, for example, has been called "perhaps the world's most famous fossil." (1990)[2389] Indubitably, this is because so few fossils can be plausibly described as "transitionals" that evolutionists have repeat-

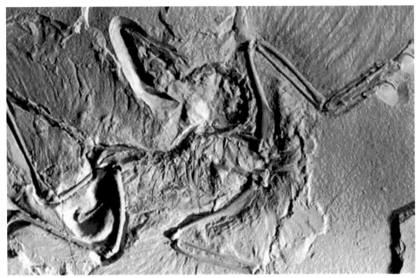

© Robert Ainsworth / Fotolia

Archaeopteryx

edly peddled *Archaeopteryx* as an example of such.

Archaeopteryx is alleged to show a link between dinosaurs and birds, but after touting it as "a prime example of an anatomical intermediate between major groups," Eldredge admits:

> We might suppose that a creature intermediate between a bipedal carnivorous dinosaur and a house sparrow would somehow be intermediate in all respects. That's not what we see at all: *Archae-opteryx* has some advanced features of birds, and some primitive retentions—anatomical features held over from its reptilian ances-try. *Archaeopteryx* had the wings and feathers of a bird, but the face and tail of a carnivorous dinosaur. *Archaeopteryx* is a melange of reptilian and bird features. It is a *mosaic* rather than a blend.... (1991)[2390]

Now we see exactly what Gould and Eldredge meant when they acknowledged there is "no evidence" in the fossil record for "smooth inter-mediates" between body plans and specifically added that "curious mosaics like *Archaeopteryx* do not count." This is critical because it is smooth inter-

mediates—not mosaics—that are needed to substantiate the theory of evolution. From the standpoint of proving evolution, there is nothing intermediate about a creature with fully developed body parts. Almost by definition,
true intermediates should demonstrate that body parts changed from one
into another. Yet, in the words of Alan Feduccia, a prominent evolutionary
biologist who specializes in birds,[2391] *Archaeopteryx* possessed:

> the feathers of modern birds, unchanged in structural detail over
> 150 million years of evolution, including microstructure, like reg-
> ular spacing of barbs throughout the feather's length and clear
> impressions of barbules. (1993)[2392] [2393]

> In fact, if you compared the claws of a wood creeper [a living bird]
> with the [wing] claws of *Archaeopteryx*, you would be hard pressed
> to tell them apart. They are virtually identical. (1993)[2394]

Are we supposed to believe that the reptile scales of a dinosaur
morphed into the fully developed feathers of *Archaeopteryx* without leaving a trace of the transition in the fossil record and, furthermore, that these
feathers then remained exactly the same for the next 150,000,000 years
down to the present day? How can anyone reasonably claim such a far-
fetched scenario qualifies as an example of an intermediate? As explained
in a 1974 college textbook:

> It seems, from the complex construction of feathers, that their
> evolution from reptilian scales would have required an immense
> period of time and involved a series of intermediate struc-
> tures. So far, the fossil record does not bear out that supposition.
> (1974)[2395]

Likewise, in a 1996 paper published in the *Journal of Evolutionary Biology*, Alan H. Brush of the Department of Physiology and Neurobiology
at the University of Connecticut noted the dissimilarities between reptile
scales and feathers are significant in every conceivable respect "from gene
structure and organization, to development, morphogenesis and tissue organization." He also wrote:

> There are no recognized intermediates in the fossil record and our
> concepts of any protofeather [ancestral feather] are speculative....
> Paradoxically there is no fossil evidence for either scales or a horny
> bill in *Archaeopteryx*. The earliest feathers in the fossil record are
> modern in every respect. (1996)[2396]

It will surely be claimed that the quote above is no longer accurate, and we will address this in a moment, but it's not just feathers that distinguish *Archaeopteryx* from dinosaurs, although you would never know this from listening to some evolutionists. As George Gaylord Simpson explained when displaying a widely used illustration of *Archaeopteryx* alongside a type of dinosaur claimed to be related to it: "Although the comparison clearly shows a relationship, it minimizes or omits altogether almost all the avian [bird] pieces of the mosaic." (1983)[2397] [2398] Alan Feduccia was a bit more direct about such misleading representations:

> Paleontologists have tried to turn *Archaeopteryx* into an earth-bound, feathered dinosaur. But it's not. It is a bird, a perching bird. And no amount of 'paleobabble' is going to change that. (1993)[2399]

Regrettably, the paleobabble only intensified from here because in 1996, it was announced that a farmer in China had found a dinosaur fossil "covered with feather-like structures."[2400] [2401] [2402] Although the primary scientific paper on this fossil warned that "much more work needs to be done to prove" these structures have any relationship to feathers, (1998)[2403] this creature, dubbed *Sinosauropteryx*, was exalted by some as a "feathered dinosaur," (1999)[2404] (2004)[2405] and swiftly in its wake came a wide assortment of "feathered dinosaur" claims, mostly based upon fossils from an "impoverished region of northeastern China."[2406] [2407] [2408] [2409] [2410] [2411] [2412] [2413]

Eager to believe a missing link had been found, many paid no heed to the fact that shortly after *Sinosauropteryx* was unveiled, roughly half a dozen paleontologists examined it and concluded the "feather-like structures" have nothing to do with feathers but are the tattered remains of collagen, a fibrous substance that is a major component of skin and many other animal body parts. (1997)[2414] [2415] [2416] Ten years later in 2007, this was borne out in a paper published in the *Proceedings of the Royal Society*, in which it was reported that microscopic examination of these structures found "a striking similarity to the structure and levels of organization of dermal collagen." As the paper explicitly declares: "The proposal that these fibers are protofeathers is dismissed."[2417] [2418]

Yet the "feathered dinosaurs" were already on the rampage, and just like in *Jurassic Park*, there was no putting them back in the coop. So it was that in November 1999, *National Geographic* published an article about a new discovery from northeastern China named *Archaeoraptor*. It was hailed as "a true missing link in the complex chain that connects dinosaurs to birds," "exactly what scientists would expect to find in dinosaurs experimenting with flight," "the best evidence since *Archaeopteryx* that birds did, in fact, evolve from certain types of carnivorous dinosaurs," and in the words of renowned dinosaur expert Philip J. Currie, "the first dinosaur that was capable of flying."[2419] [2420]

A few months later, *Archaeoraptor* was proven to be a forgery constructed by a Chinese farmer with homemade cement. Simply put, he glued together rock slabs containing the remains of two different creatures.[2421] [2422] As ridiculous as this is, the larger story is even more outrageous because the researchers who examined the fossil knew parts of it had been cobbled together but frivolously claimed they had come from the same creature.[2423] Worse still, even though the *National Geographic* researchers knew the fossil had been tampered with, none of this was mentioned in the *National Geographic* article.[2424] On top of this, a paper written about the fossil by the *National Geographic* research team was rejected by the prestigious journal *Science*—not because the peer review process had uncovered the fraud but because the reviewers wanted more proof of the fossil's birdlike qualities.[2425] Ironically, the main body of the fossil is, in fact, a bird.[2426] So in sum, the paper was rejected for the primary element of it that was actually based on reality.

Given all the events that resulted from a forgery made by a peasant farmer in his one-room home,[2427] can we possibly imagine what could happen with the product of a sophisticated forger? The incentives for this kind of work are considerable, and as is typical with illicit gain, there is no shortage of people willing to participate. The *Archaeoraptor* profiteers included the farmer, a dealer, and, quite possibly, bribed government officials who shepherded the fossil's export to the U.S. where it sold for $80,000.[2428] A scientist at the Institute of Vertebrate Palaeontology in Beijing provides some insight into the situation:

> Farmers or dealers can make a much bigger profit if they've got the fossil of a complete animal. They take a damaged specimen and add the parts from other animals to make a new one which looks complete. In one place I saw them putting all the bits from a dinosaur's leg into a box, just like a box of machine spare parts so that they could add them to different fossils. (2002)[2429]

In addition to bonding parts together, another method of forgery discovered in fossils from this region includes carving added features. Also, it is clear some of the forgers are not crafting their designs based upon artistic whim but are employing knowledge of paleontological data.[2430] These facts buttress these words of Alan Feduccia:

> *Archaeoraptor* is just the tip of the iceberg. There are scores of fake fossils out there, and they have cast a dark shadow over the whole field. When you go to these fossil shows, it's difficult to tell which ones are faked and which ones are not. I have heard that there is a fake-fossil factory in northeastern China, in Liaoning Province, near the deposits where many of these recent alleged feathered dinosaurs were found. (2003)[2431]

Adding to the suspect circumstances, these same deposits contain birds bearing feathers that are "essentially identical" to those on some of the "feathered dinosaur" fossils.[2432] In 1996, more than 130 years after *Archaeopteryx* was discovered,[2433] Alan H. Brush wrote there "are no recognized intermediates [between reptile scales and feathers] in the fossil record and our concepts of any protofeather are speculative." Yet, a mere six years later in 2002, he and a coauthor voiced the polar opposite conclusion, citing seven "startling new paleontological discoveries of primitive feathers" in dinosaurs—every one of them from the same province of northeastern China.[2434]

So let's get this straight. After more than 130 years with no evidence of such feathers in the fossil deposits of the entire world, all of a sudden, seven such "startling" discoveries turn up in a locality that happens to be "home to a highly developed faking industry." (2002)[2435] Am I asserting every one of these fossils is a forgery? No, but I am asserting that between the suspect circumstances associated with these finds and the manner in which evolutionary presumptions have been shown to disfigure reality, the evidence of any so-called transition between reptile scales and bird feathers is extremely doubtful on multiple levels. As Alan Feduccia and two coauthors sarcastically declare in a paper contesting the notion that any feathered dinosaur ever existed:

> [I]f a modern kiwi were discovered in the [lake] deposits of the Early Cretaceous [≈100–146 mya] of China, it would most assuredly be considered a theropod dinosaur, illustrating an early stage in the evolution of flight from the ground up, and adorned with protofeathers and all stages of feather evolution. (2005)[2436]

Before it was publicly exposed as a forgery, *Archaeoraptor*, along with models of other "feathered dinosaurs," sat on display at the National Geographic museum in Washington D.C. where they were viewed by 110,000 visitors, most of them children.[2437] During this period, a scientist at the Smithsonian Institution sent a letter to the most prominent scientist at the National Geographic Society, stating that with the publication of the article concerning *Archaeoraptor*:

> *National Geographic* has reached an all-time low for engaging in sensationalistic, unsubstantiated, tabloid journalism. …
>
> The hype about feathered dinosaurs in the exhibit currently on display at the National Geographic Society is even worse, and makes the spurious claim that there is strong evidence that a wide variety of carnivorous dinosaurs had feathers. A model of the undisputed dinosaur Deinonychus and illustrations of baby tyrannosaurs are shown clad in feathers, all of which is simply imaginary and has no place outside of science fiction.
>
> The idea of feathered dinosaurs and the [dinosaur] origin of birds is being actively promulgated by a cadre of zealous scientists acting in concert with certain editors at *Nature* and *National Geographic* who themselves have become outspoken and highly biased proselytizers of the faith. Truth and careful scientific weighing of evidence have been among the first casualties in their program, which is now fast becoming one of the grander scientific hoaxes of our age…. (1999)[2438]

Such has been the case with any alleged transitional fossil I have investigated. The evidence shows blunt indications the claims are driven by ideology instead of science. Admittedly, I have not taken the time to thoroughly weigh each claim because, for the purpose of answering the question before us, there is no need. If evolution has any basis in reality, the fossil record should be teeming with such forms, but as demonstrated by numerous corroborating quotes and evidences provided by evolutionists from Darwin to the present day, there are very few, and it would be charitable to call the most highly touted of these doubtful. Such a state of affairs does not come even close to what would be the case if every living creature and plant descended from a microbe.

IT'S JUST TOO MUCH

Given what science has revealed about the fossil record, one can easily understand why Oxford zoologist Mark Ridley would write that "no real evolutionist … uses the fossil record as evidence in favor of the theory of evolution as opposed to special creation." (1981)[2439] The same applies to any discipline or line of evidence one cares to examine. For the sake of argument over the last four chapters, we have:

- put aside that the big bang theory has been capriciously and repeatedly altered to keep it from collapsing in the face of new evidence, and yet, is still rife with major problems and discrepancies;
- swallowed the discredited doctrine of spontaneous generation and accepted that organisms far more complicated than a computer arose from non-living matter, despite 140 years of experiments that have never even remotely produced such results;
- pretended that a theory requiring the natural formation of countless biochemical and cellular systems may be true, despite the failure to find a single example of such in the entire kingdom of nature, and
- overlooked the fact that what have been heralded for more than a century as leading "proofs" of evolution such as embryology and vestigial organs are based upon misinformation, distortions, and outright fraud.

And these are just a few of the evolutionary fallacies exposed over the past four chapters. Far too much space would be required to summarize each fallacy, but suffice it to say the obvious: at every step of the way, the theory of evolution is at odds with the scientific evidence. One might be able to live with a few of these conflicts and still accept the theory, but to brush them all aside defies reason. It's just too much for any informed and unprejudiced person to abide.

So why is it that so many intelligent people with advanced degrees accept the theory of evolution? Simple: they are either under misconceptions like those detailed herein, and/or they won't even consider the idea that God created the universe as the Bible affirms. As various evolutionists quoted in Chapter 6 declared: "[W]e are compelled by our calling to insist at all times on strictly naturalistic explanations,"[2440] and, "Even if all the data point to an intelligent designer, such an hypothesis is excluded from science because it is not naturalistic."[2441] In sum, they wear an intellectual straightjacket sewn of the cloth "that materialism is absolute, for we cannot allow a Divine Foot in the door."[2442] All of this goes to show that even genius will lead to darkness if it begins with false assumptions.

RELATIVITY AND THE BEGINNING OF TIME

If the universe and the astonishing things within it cannot rationally be ascribed to natural causes, one has no choice but to acknowledge there must be something more. Hence, coming to the realization stated in the Bible that God is evident from the things He has made,[2443] many have wondered, "What came before God?"

By definition, there is no such thing as "before God." We reside in the dimensions of time and space, but God, who created this system, exists independently of it. As the Bible explains, God "inhabits eternity" and exists from "everlasting to everlasting."[2444] In contrast, the dimensions of time and space, regardless of whatever our intuitions may be, have not always existed. This is not mere philosophy but an implication of Einstein's theory of general relativity. As explained by Jim Al-Khalili, a physicist with a Ph.D. in theoretical nuclear physics from the University of Surrey, U.K.:

> If Einstein's general theory of relativity is correct, and we are confident that it is, then the Big Bang not only marked the birth of the Universe but the beginning of time itself. Asking questions about what happened *before* the Big Bang necessitates having time to imbed the word 'before' in. Since there simply was no time before the Big Bang, the question doesn't make sense. (1999)[2445]

More specifically, in the words of acclaimed physicist Stephen Hawking:

> Exactly what Einstein's general theory of relativity predicted in these situations remained unclear until Roger Penrose and I proved a number of theorems. These ... implied that there would be singularities, places where space-time had a beginning or an end. (1991)[2446]
>
> Not surprisingly, many scientists were unhappy with this conclusion. There were thus several attempts to avoid the conclusion that there must have been a big bang singularity and hence a beginning of time. ...
>
> ... We showed that if general relativity is correct, any reasonable model of the universe must start with a singularity. This would mean that science could predict that the universe must have had a beginning, but that it could not predict how the universe *should* begin: For that, one would have to appeal to God. (1987)[2447]

For those who find it difficult to comprehend how reality can consist of more than what we can perceive with our five senses and scientific instruments, a striking allegory can be found in a brilliant fictional work used by mathematicians to illustrate the concept of dimensions.[2448] In this classic book entitled *Flatland*, the author describes a relationship between a being that exists in three spatial dimensions (as we do) and a being that exists in only two dimensions.[2449] This illustration allows us to see that the three-dimensional being, by virtue of his existence in an added dimension, has far more perspective, can reveal himself at will to the two-dimensional being, and can perform deeds the two-dimensional being regards as super-natural, which includes lifting him out of his limited world and into the higher dimension.

I don't know that this analogy accurately applies to God, but it allows us to understand that our perception of reality may in fact be very limited. What we would regard as impossible or miraculous could be effortlessly done by a being that transcends the temporal dimensions of space and time in which we live. In Chapter 2, I asserted that "the most compelling scientific evidence for the reality of miracles is the very existence of our universe and the marvels found within it." With all we have seen since then, it should be obvious why.

CHAPTER 10
AUTHENTIC CHRISTIANITY

While I was in the midst of writing this book, a friend posed the following question to me: "What is the most profound thing you have learned from your research so far?" I don't recall the exact words, but my reply was something like this:

> If God decided to, He could reveal himself in such a way that no one could ever doubt His existence. Yet, with a few notable exceptions, such as Moses, normally He chooses not to, and I think the reason is that we would never develop a sincere relationship with Him under such circumstances. How could we develop genuine affection for a completely righteous, all-powerful, and all-knowing being if His existence were always inescapably clear? God desires true love from us, not the kind that is born of fear and brownnosing. Thus, He gives us ample evidence to conclude He exists but normally stops short of undeniable proof so those who embrace Him will do so of their own free will.[2450]

Somewhat similarly, a clergyman by the name of Daniel Wilson said the following in the early 19th century:

> Christianity does not profess to convince the perverse and headstrong, to bring irresistible evidence to the daring and profane, to vanquish the proud scorner, and afford evidences from which the careless and perverse cannot possibly escape. This might go to destroy man's responsibility. All that Christianity professes, is to propose such evidences as may satisfy the meek, the tractable, the candid, the serious inquirer.[2451]

I couldn't agree more. The evidences supporting the accuracy of the Bible are abundant and compelling, but they are not enough to overwhelm a closed mind. And herein lies a great irony, for what closes many minds to Christianity is Christians, or at least those who claim to be. In the words of Australian scientist Andrew Ruys:

> I am obviously not alone in the viewpoint that the more I see

through science, the more I see God. In my experience, it is not usually science that causes people to make a decision to become an atheist. Usually, it is a subjective reason, for example, a negative response to a repressively religious church, school, or upbringing.[2452]

Real Christians

The impression some people have of Christianity is distorted because they equate it with every person who has ever called themselves a Christian. This mindset is demonstrably groundless, especially in a culture where 68% of people describe themselves as "committed Christians"[2453] yet only 16% claim to base their moral decisions on the Bible.[2454] In other words, three out of four Americans who profess to be Christians don't even claim to consult the words of Christ when making ethical decisions. Moreover, this statistic doesn't even begin to account for the duplicitous actions of many Bible toters. All of this amounts to an enormous disconnect that echoes and fulfills these words of Jesus recorded in the Book of Matthew:

> Not everyone who says to me, "Lord, Lord," will enter the kingdom of heaven, but only he who does the will of my Father who is in heaven. Many will say to me on that day, "Lord, Lord, did we not prophesy in your name, and in your name drive out demons and perform many miracles?" Then I will tell them plainly, "I never knew you. Away from me, you evildoers!"[2455]

Given the above, to pin on Christianity the wrongs of everyone who has ever professed to be a Christian is senseless. Even more illogical is to make blanket statements about the misdeeds of "religion."[2456] Should pacifists of one faith be categorized with militants of another because they are both "religious"? It is ironic that those who make such sweeping generalizations don't apply the same standard to everyone who has expressed anti-Christian or atheistic views. For if they did, all atheists and critics of Christianity would find themselves in the company of Hitler,[2457] [2458] Lenin,[2459] Stalin,[2460] Mao Tse-Tung,[2461] and a horde of other malefactors who have collectively murdered untold millions of innocent people.

"Don't Know Much About History"[2462]

Many mistaken notions about Christianity have their roots in contextual shortcomings that pervade books, newspapers, and movies. For example, most of us have been exposed to a good deal of information about the Holocaust, but how many of us know about these words of Albert Einstein reported in *Time* magazine in 1940? Read them carefully, for they are bursting with what is missing from many historical accounts of this episode in history:

> Being a lover of freedom, when the revolution came in Germany, I looked to the universities to defend it, knowing that they had always boasted of their devotion to the cause of truth; but, no, the universities immediately were silenced. Then I looked to the great editors of the newspapers whose flaming editorials in days gone by had proclaimed their love of freedom; but they, like the universities, were silenced in a few short weeks. … Only the Church stood squarely across the path of Hitler's campaign for suppressing truth. I never had any special interest in the Church before, but now I feel a great affection and admiration because the Church alone has had the courage and persistence to stand for intellectual truth and moral freedom. I am forced thus to confess that what I once despised I now praise unreservedly.[2463] [2464]

Einstein was in an exceptional position to make such an observation. Besides the fact that he was a German citizen of Jewish descent, he was also a renowned intellectual with privileged access to universities and the news media. Moreover, as an agnostic who previously "despised" the Church, Einstein had no ideological motivation to make such a statement.[2465] [2466]

Some of the greatest evidences for the veracity of the Bible are found in the lives of those who follow the principles in it. One of these principles, by the way, includes not advertising one's own good deeds.[2467] Couple this with a press corps and entertainment industry that is often contemptuous of Biblical faith[2468] [2469] (which, incidentally, is another fulfillment of Christ's words[2470]), and it is no wonder that some people have an ill-informed and negative view of Christianity. Even those who are not so easily duped by such propaganda rarely have the time to read the primary sources that truly enlighten the events of present and past, such as these words penned by Booker T. Washington:

> If no other consideration had convinced me of the value of a Christian life, the Christlike work which the Church of all denominations in America has done during the last thirty-five years for the elevation of the black man would have made me a Christian.[2471]

In order to gain an accurate perspective of any belief system, we must look at the actions of people who faithfully apply the principles it espouses. This is why it is so important to interpret the Bible accurately. Some will claim you can interpret the Bible to mean anything you would like it to mean. From a person who read the Bible cover to cover looking for flaws and contradictions during a time in my life when I thought it was fiction, I can affirm this is not the case. On some issues, there are legitimate grounds for differing interpretations, but for the most part, this is only the case if one takes the Bible out of context. It is vital to understand this because people construe all sorts of bogus interpretations from the Bible to justify their personal agendas.

Accurately Interpreting the Bible

Because the Bible was written long ago in foreign languages, grasping the meaning of it sometimes requires more than just carefully reading it. Hence, in the spirit of Booker T. Washington, the topic of slavery offers a great example to familiarize ourselves with four principles for accurately interpreting the Bible. These are listed here and explanations of how to employ them follow:

1) Be intellectually honest.

2) Acquire a broad view.

3) Be aware of translational issues.

4) Consider the setting and historical context.

Intellectual Honesty

Given the following passages, one might think the matter of slavery would be a very simple issue:

> [Jesus said] in everything, do to others what you would have them do to you, for this sums up the Law and the Prophets.[2472]

> [Paul said] Love your neighbor as yourself. Love does no harm to its neighbor. Therefore love is the fulfillment of the law.[2473]

How could anyone possibly reconcile the practice of slavery with these words? Yet the historical record is clear that a number of "Christians" attempted to do so. In 1856, a clergyman insisted that the words of Jesus above are compatible with slavery because much the same thought appears in the Old Testament, yet it supports slavery.[2474] As we will see in a moment, this is misleading on several accounts. For now, however, the first and most important principle to be learned in accurately interpreting the Bible (or anything else for that matter) is to be truthful with ourselves. Exploiting technical loopholes to avoid the obvious is a hallmark of intellectual dishonesty. There is simply no rational way to square the Bible passages above with abducting innocent people and forcing them to work under threat of corporal punishment. With this, we could consider the issue of slavery settled, but we're going pursue it further because much can be learned from it.

A Broad View

The second principle for interpreting the Bible accurately is to consider all the pertinent passages on a given topic. Some parts of the Bible are open to varying interpretations if taken in isolation, but when you bring together all of the relevant material to arrive at a complete picture, the point is almost always clear, particularly when it comes to ethical matters. For instance, it is true the Old Testament sanctions "slavery," but vast differences exist between this and the form of slavery with which we are familiar. Let's start by looking at a passage that is clearly supportive of "slavery":[2475] [2476]

> If one of your countrymen becomes poor among you and sells himself to you, do not make him work as a slave. … Your male and female slaves are to come from the nations around you; from them you may buy slaves. You may also buy some of the temporary residents living among you and members of their clans born in your country, and they will become your property. You can will them to your children as inherited property and can make them slaves for life….[2477]

The first difference to note is that people sold themselves (not others) into servitude. In the Old Testament, the type of "slavery" sanctioned is not forcible. It is something a person voluntarily enters into in exchange for money or a service. This is amply proven by the following passages:

> Anyone who kidnaps another and either sells him or still has him when he is caught must be put to death.[2478]

> If a slave has taken refuge with you, do not hand him over to his master. Let him live among you wherever he likes and in whatever town he chooses. Do not oppress him.[2479]

An exception to this is in cases of criminal conduct. A person caught stealing was required to make restitution to the victim, and if he could not, he was sold and the money given to the victim:

> A thief must certainly make restitution, but if he has nothing, he must be sold to pay for his theft.[2480]

This certainly has similarities with our modern penal codes in which criminals are sometimes sentenced to "imprisonment with hard labor."[2481] We may recoil at the thought of selling oneself into servitude, but there are parallels between this and contracts that are made every day. Such contracts are typically for a set length of time, but this is not always the case, and in certain situations, the U.S. government, for example, can legally compel someone to serve in the military for an indefinite period.[2482] [2483]

In summary, the lesson to be learned is to consider the Bible as a whole instead of selectively picking from it. Many study Bibles contain references and commentaries that enable one to acquire this broad view.

Translational Issues

Interpretation principle number three is to be aware of translational issues. This does not mean you need to learn another language, because the website blueletterbible.org makes this type of investigation relatively simple. For instance, another major difference between the type of "slavery" sanctioned in the Old Testament and the kind we are acquainted with is that harsh treatment was prohibited. If a master knocked out the tooth of a slave, the slave was to be freed:

> If a man hits a manservant or maidservant in the eye and destroys it, he must let the servant go free to compensate for the eye. And if he knocks out the tooth of a manservant or maidservant, he must let the servant go free to compensate for the tooth.[2484]

Wait a moment. This passage does not refers to slaves, but to servants. Are they one and the same? In this case, the answer is probably yes. By using the resources at blueletterbible.com, we see that the Hebrew word translated as "slave" in the foregoing passages is "ebed."[2485] The same word is used directly above but is translated as "servant." As was explained in Chapter 5 and demonstrated in other parts of this book, no Bible translation is perfect. Beyond the fact that there are many things we don't fully understand about ancient Hebrew, it is simply not plausible to convey every nuance of one language into another. In fact, it has been convincingly argued there isn't even a word in ancient Hebrew that truly corresponds to the modern meaning of the word "slave."[2486]

Setting and Historical Context

The fourth and last principle we will cover is to consider the setting and historical context. Simply quoting from the Bible can be misleading if we don't understand who is speaking, who is being spoken to, and the surrounding circumstances. Again, you don't need to be a scholar to do this. Any good study Bible will contain an ample amount of such information.

The New Testament is set in the Roman Empire at a time when slavery was widespread and cruelty to slaves was at or near its peak. Just to entertain their guests, Roman masters sometimes killed their slaves or hacked their limbs off.[2487] During this time, the apostle Paul sent to churches and individuals various letters that became a significant part of the New Testament. In these letters, Paul makes several references to slavery, such as the following:

> Slaves, obey your earthly masters with respect and fear, and with sincerity of heart, just as you would obey Christ. Obey them not only to win their favor when their eye is on you, but like slaves of Christ, doing the will of God from your heart. Serve wholeheartedly, as if you were serving the Lord, not men, because you know that the Lord will reward everyone for whatever good he does, whether he is slave or free. And masters, treat your slaves in the same way. Do not threaten them, since you know that he who is both their Master and yours is in heaven, and there is no favoritism with him.[2488]

Since slaves are told to be obedient to their masters, at first glance one might think this passage is supportive of slavery. However, masters are instructed not to threaten their slaves and to treat them "in the same way" slaves are to treat their masters. How could slavery exist under such guidelines? The answer is it can't. So why doesn't Paul directly tell masters to set their slaves free?

This is where historical context becomes important. The Roman historian Suetonius recorded that in the era when Jesus was born, the Emperor Augustus enacted "many obstacles to either the partial or complete emancipation of slaves."[2489] [2490] [2491] Thus, instead of calling for the release of slaves that would have resulted in a fruitless conflict with the Roman Empire, Paul undercut the institution of slavery by advancing values irreconcilable with it. Some have tried to put a different slant on Paul's words by interpreting them from the viewpoint that slavery was a benefit to the slaves, but we know Paul didn't think this way because he included "slave traders" in a list of "ungodly and sinful" people.[2492] Furthermore, he wrote the following to believers in the city of Corinth, Greece:

> Were you a slave when you were called? Don't let it trouble you—although if you can gain your freedom, do so.[2493]

Further insight is found in a letter written by Paul to a Christian named Philemon, whose slave apparently ran away, met Paul, and became a Christian. Paul sent this slave (named Onesimus) back to Philemon with a letter stating:

> I am sending him—who is my very heart—back to you. … Perhaps the reason he was separated from you for a little while was that you might have him back for good—no longer as a slave, but better than a slave, as a dear brother. He is very dear to me but even dearer to you, both as a man and as a brother in the Lord. So if you consider me a partner, welcome him as you would welcome me. If he has done you any wrong or owes you anything, charge it to me. … Confident of your obedience, I write to you, knowing that you will do even more than I ask.[2494]

Here, Paul, in compliance with Roman law, sends a slave back to his master but tells the master to welcome this person as he would welcome Paul himself, "no longer as a slave," but as a "dear brother." Then he boldly enjoins the master to do "even more than I ask." These are obviously instructions to set Onesimus free, or at the very least treat him as such if there were legal impediments to actually freeing him. Also, note Paul uses the term "brother." This carries profound implications that ring throughout his writings:

> You are all sons of God through faith in Christ Jesus…. There is neither Jew nor Greek, slave nor free, male nor female, for you are all one in Christ Jesus.[2495]

A minister named Charles Elliot eloquently articulates this point in a book published two years before the start of the U.S. Civil War:

> To apply the terms brethren and sisters to slaves, initiates a new element into the subject unknown to all slave laws, and all slavery principles. In the West Indies the pro-slavery men … ridiculed the idea of brothers and sisters among the missionary Churches. They

asked, "Can you make your negroes Christians, and use the words *dear brother* or *sister*, to those you hold in bondage? They would conceive themselves, by possibility, put on a level with yourselves, and the chains of slavery would be broken."[2496]

He also provides a poignant summary of the New Testament's response to the Roman slave system:

To have preached the emancipation of slaves, by the apostles, would have been the same as to attempt an overthrow of the Roman Government. And this civil emancipation would not strike at the root of the evil. Our Lord and his apostles, therefore, went to the source of the evil, by preaching the Gospel to both slaves and masters; so that, in carrying out the moral principles of our holy religion, and a moral practice under it, the great moral evils of the world were undermined.[2497]

Indeed, when one studies the abolition movement, we see that most of the leading figures who made immense personal sacrifices for this cause were dedicated Christians acting in accordance with Biblical principles. This includes William Wilberforce, Thomas Clarkson, Granville Sharp, Frederick Douglass, William Lloyd Garrison, John Newton, James Ramsay, James Stephen, Elizabeth Heyrick, and many others.[2498] [2499] To summarize this section in the same words that introduced it:

In order to gain an accurate perspective of any belief system, we must look at the actions of people who faithfully apply the principles it espouses. This is why it is so important to interpret the Bible accurately.

THE SUBJECTIVE NATURE OF MORALS AND VALUES

There is one more point we will contend with in examining why some are close-minded to Christianity. There are people, and I am one of them, who understand exactly what the Bible says about certain subjects and simply don't like it. For example, Jesus said, "Love your enemies, bless them that curse you, do good to them that hate you, and pray for them which despitefully use you...."[2500] To be blunt, I would have much preferred if Jesus had said to give such people a good beating. It's very difficult to love people like that, and from my own personal perspective, I don't feel it's justified. However, I realize these are just emotions, and emotions don't define what is right and what is wrong.

Regrettably, too few of us realize that some of our deepest values may rest on completely arbitrary grounds. We all have views about what constitutes right and wrong, but if pressed to give a rational explanation for these beliefs, could we?

Consider this forceful example: With the exception of a few unhinged or ill-informed individuals, most everyone would agree with me that Hitler committed many atrocious wrongs. Yet, if Hitler were here to debate that point, there can be little doubt he would argue that what he did was good. And if you think about it, on what objective basis could we prove otherwise? If you challenged him face to face and told him he was despicable for brutally slaughtering all those innocent people, he could simply reply: "There is nothing wrong in what I did. In fact, it was for the common good of the human race." What, then, could you say to him? To answer this requires a hard look at the nature of morality.

Everyone judges his or her own values to be right. Even the person who says, "To each his own," imposes that judgment upon others. Who are we to tell Hitler that he can do whatever he wants until he harms someone else? Truth be told, we deem our morals superior to his. Yet, if there is no such thing as absolute morality and it's all a matter of personal preference as moral relativists say, then one person's preference is as good as another's, including Hitler's.

In response to Hitler's appalling views, a Christian could rationally reply that God, the ultimate judge of right and wrong, forbids the shedding of innocent blood. Believe me, I'm under no illusion that Hitler would do anything other than scoff at this, but it is not the mind of Hitler we need to satisfy—it is our own. However, what could the atheist say? "Moral decisions need to be based on logic and experience, and I have learned through these that murder is wrong." Hitler would undoubtedly respond: "Survival of the fittest is a fact of nature, and what I did is perfectly consistent with this fact. By ridding society of inferior beings, I improved the human race. Logic and experience dictate that building a master race is ethical, not immoral."

Then what could the atheist say? Think about it. If you'd like, take some time to read atheist writings about morality, many of which you can find on the Internet. You'll find there is no reply of any substance. If we are merely the products of evolution, why should anyone have a problem with the strong killing the weak? The weak are a drain on society. Worse yet, they pollute the human gene pool and slow the advancement of mankind. Note that I am not voicing hypothetical ideas through Hitler's mouth.

This is exactly how he reasoned, and it is entirely consistent with the view that there is no God. For instance, on October 10, 1941, Hitler stated:

> Today war is nothing but a struggle for the riches of nature. By virtue of an inherent law, these riches belong to him who conquers them. … That's in accordance with the laws of nature. By means of the struggle, the elites are continually renewed. The law of selection justifies this incessant struggle, by allowing the survival of the fittest. Christianity is a rebellion against natural law, a protest against nature. Taken to its logical extreme, Christianity would mean the systematic cultivation of the human failure.[2501]

In accordance with this, Hitler ordered the production of a propaganda film entitled *Victim of the Past* and required it to be shown in all German movie theaters. In it, footage of a psychiatric institution is shown while a narrator declares:

> Wherever fate puts us, whatever station we must occupy, only the strong will prevail in the end. Everything in the natural world that is weak for life will ineluctably be destroyed. In the last few decades, mankind has sinned terribly against the law of natural selection. We haven't just maintained life unworthy of life; we have even allowed it to multiply! The descendants of these sick people look like this![2502]

Fewer than three years after this film was released, German mental hospitals began gassing to death thousands of innocent people.[2503] Sadly, Hitler's morals were more grounded in logic than the morals of genuinely compassionate atheists. Don't get me wrong. I am not impugning the morals of atheists. I have tremendous respect, love, and appreciation for many people who don't believe in God. In fact, from what I can see, some of them live their lives more consistently with the Bible than some people who walk around with one. The point of fact, however, is that apart from a supreme Lawmaker, no objective basis exists for any morality except the law of the jungle: "The strong survive and the weak be banned."[2504] As Richard Dawkins, a leading atheist, has written:

> This is one of the hardest lessons for humans to learn. We cannot admit that things might be neither good nor evil, neither cruel nor kind, but simply callous—indifferent to all suffering, lacking all purpose.[2505]

Without God, Dawkins would be 100% correct. Like many people in our society, I grew up with values mostly identical to those in the Bible: be honest, be compassionate, do not steal, etc. However, by no one's prerogative but my own, some of my values differed from Biblical values. This is to be expected given that different people embrace a wide range of morals and even alter their own over the course of time. Then what, if any, morals are rationally justifiable? Logic demands that if God truly exists, His values are the only objective standards of right and wrong. Thus, to reject the Bible because we don't care for some of its moral guidelines is an exercise in irrationality. As explicated by Johannes Kepler, the brilliant physicist and person singled out by Immanuel Kant as "the most acute thinker ever born":[2506]

> If the intellect has agreed to contemplate what God had made, it also agrees to do what God has bid.[2507]

For me, being a Christian is a triumph of intellect over emotion. Christianity is not a philosophy I liked and bought into. It is a truth I recognized and embraced. This is not to take any credit for myself or to discount the Biblical truth that it is God who draws us to Him,[2508] but rather to say He can do this through our heads as well as our hearts.

The Gospel

With this, let's examine a series of Biblical principles central to the Christian faith, often referred to as "The Gospel." A truth foundational to all of these principles is that God loves us and desires that we love Him and each other:

> "Yes, I have loved you with an everlasting love; Therefore with lovingkindness I have drawn you."[2509]

> Jesus said to him, "You shall love the LORD your God with all your heart, with all your soul, and with all your mind. This is the first and great commandment. And the second is like it: You shall love your neighbor as yourself."[2510]

Yet, every person has sinned, and this leads to separation from God and to death:

> But your iniquities have separated you from your God; And your sins have hidden His face from you, So that He will not hear.[2511]

> If we say that we have no sin, we deceive ourselves, and the truth

is not in us.[2512]

For the wages of sin *is* death….[2513]

Why, then, doesn't God immediately do away with all evil? The answer, if we are honest with ourselves, is that none of us would be left standing. God, however, does not want any of us to perish:

Even so, it is not the will of your Father which is in heaven, that one of these little ones should perish.[2514]

The Lord is not slow in keeping his promise, as some understand slowness. He is patient with you, not wanting anyone to perish, but everyone to come to repentance.[2515]

Hence, God sent his Son to live among us and open the door to eternal life despite our shortcomings:

This is love: not that we loved God, but that he loved us and sent his Son as an atoning sacrifice for our sins.[2516]

For God so loved the world that He gave His only begotten Son, that whoever believes in Him should not perish but have everlasting life.[2517]

And this is eternal life, that they may know You, the only true God, and Jesus Christ whom You have sent.[2518]

With this, we must realize no one can earn eternal life through good deeds. It is simply a gift of God. One of the most prevalent misconceptions concerning eternity is summed up in this statement of Muhammad Ali: "One day we're all going to die, and God's going to judge us, [our] good and bad deeds. [If the] bad outweighs the good, you go to hell; if the good outweighs the bad, you go to heaven."[2519] Contrast this with what the Bible states:

[A] man is not justified by observing the law, but by faith in Jesus Christ.[2520]

For by grace are ye saved through faith; and that not of yourselves: it is the gift of God….[2521]

[No one] can keep alive his own soul.[2522]

Not by works of righteousness which we have done, but according to his mercy he saved us….[2523]

No one but Christ has lived a life so completely free of sin that he or she can stand before God and honestly say, "I deserve to be in heaven. I've earned it." We cannot earn our way into heaven. It is simply a gift. One could say it is a gift with strings because Jesus said if you love Him you'll obey his commands, but it is really a gift with keys because the things he calls us to do bring long-term happiness.[2524] And as pointed out earlier in this chapter, real Christians will display tangible evidence of our relationship with God through our works:

> [Jesus said] Whoever has my commands and obeys them, he is the one who loves me.[2525]

> And hereby we do know that we know him, if we keep his commandments.[2526]

> Show me your faith without your works, and I will show you my faith by my works.[2527]

Again, such works do not justify us to be in the presence of God, and despite the oft-repeated view that there are "many paths to God," this is simply not the case:

> Jesus saith unto him, I am the way, the truth, and the life: no man cometh unto the Father, but by me.[2528]

> [John the Baptist said] Whoever believes in the Son has eternal life, but whoever rejects the Son will not see life, for God's wrath remains on him.[2529]

I had a real problem with this when I first read it. I thought, "That's harsh. The Bible says God is loving and fair.[2530] If that's true, why would God have wrath against someone just because he doesn't have the right beliefs?" This didn't seem loving or fair to me, but after putting some serious thought into the matter, I recognized the obvious. If in fact God willingly surrendered the life of His Son for us, is it too much of Him to ask that we recognize this?

Think about it. We can't refuse to accept God's gift and then reasonably claim God isn't loving or fair because we just don't want to take it. I can't imagine what I could say if I came before God when my time on this earth was completed and He asked, "Why didn't you accept my gift?" What possible excuse could I offer? Could I tell God it wasn't a part of my religion? Or that I knew people who called themselves Christians and didn't like them? Or what if I told God I just didn't agree with some of the moral laws in the Bible and therefore rejected it? Consider this from the perspective of a parent. How would you feel if you let your child die so that someone else could live and that person was willfully ignorant or dismissive about it?

The only legitimate answer I think anyone could offer is he or she was truly unaware—for instance, a person who has never heard about Christ or a child who was too young to understand. Although the Bible doesn't explicitly address such scenarios, passages such as the following make me fairly certain God does not hold legitimate ignorance against anyone:

> Jesus said, "For judgment I have come into this world, so that the blind will see and those who see will become blind." Some Pharisees who were with him heard him say this and asked, "What? Are we blind too?" Jesus said, "If you were blind, you would not be guilty of sin; but now that you claim you can see, your guilt remains."[2531] {For those who may question how this reconciles with Romans 1:18–21, see this note.[2532]}

GOD'S DESIRE FOR OUR LIVES

Some are reluctant to embrace God for fear they will lose their individuality, but nothing could be further from the truth. We didn't make ourselves. God did, and He knows us infinitely better than any of us can actually know ourselves. Thus, the closer we draw to God, the more like ourselves we truly will be. In the words of James Clerk Maxwell, the renowned physicist whose incredible scientific achievements were detailed in Chapter 2:

> The consciousness of the presence of God is the only guarantee for true self-knowledge. Everything else is mere fiction, fancy portraiture,—done to please one's friends or self; or to exhibit one's moral discrimination at the expense of character.[2533]

As Maxwell knew and wrote, "Man's chief end is to glorify God and to enjoy him for ever."[2534] To enjoy this eternity only requires accepting God's gift. The exact words are not critical because God can hear your heart, but these get the point across:

> Thank you for sending your Son to bear the consequences of my sins. I accept this gift in all earnestness. Please give me the faith, love, and strength to be the person You created me to be. I ask and thank You for all this in the name of Your Son Jesus Christ.

If you can sincerely say that to God, you have established a relationship with Him that runs far deeper than you can possibly imagine at this point. Thus, I highly recommend getting a good study Bible and beginning your journey in the New Testament. One of the best, in my opinion, is the *Life Application Study Bible.*[2535]

The New Testament tells us that believers should spend time together,[2536] which naturally leads to the topic of church. It is not my desire to advocate a specific type of church or denomination but rather to forewarn against "false prophets" as Jesus did:

> Beware of false prophets, which come to you in sheep's clothing, but inwardly they are ravening wolves. Ye shall know them by their fruits. Do men gather grapes of thorns, or figs of thistles? Even so every good tree bringeth forth good fruit; but a corrupt tree bringeth forth evil fruit.

Again, God sets the standards of what is good and evil, so be certain to rely on the Bible in this process. In addition to the verses already quoted in this chapter, there are others, such as 2 Timothy 3:1–7[2537] and Colossians 1:18,[2538] that are particularly important in selecting a church. However, one verse on this subject is so vital I would be remiss not to highlight it:

> [Jesus said:] By this shall all men know that ye are my disciples, if ye have love one to another.[2539]

In our culture, the word "love" has been so warped that many people don't understand what Jesus meant when He said this. The Greek word translated in this passage as "love" is "agape," and by using blueletterbible.org, we can examine exactly where and how this word is used in the Bible. When we do this, we see the love to which Christ called us is not a mushy sentiment but a commitment to sacrifice for the benefit of others. It requires generosity,[2540] humility, and patience.[2541] It demands we obey God,[2542] examine ourselves, and hold each other accountable.[2543] It compels us to place the needs of others before our own preferences and calls us to use our freedom to serve.[2544] [2545] It casts out fear and rejects lust, pride, and the love of material objects.[2546] [2547] It involves work.[2548] It is marked by pure intentions, not the desire to manipulate.[2549] It forbids us from stealing, committing adultery, and lying.[2550] It requires us to be kind while rejecting envy and evil thoughts.[2551] It demands we be wary of deceitful rhetoric and call it out for what it is.[2552] It is so abiding it cannot be shattered by trials, persecution, or even death.[2553] In fact, it may even require us to lay down our life for another, just as Christ did for us.[2554] As aptly summarized in the Book of First John:

[L]et us not love in word, neither in tongue; but in deed and in truth.[2555]

In 1858, James Clerk Maxwell wrote for his wife a poem concluding with profound words that now close this book. Read them carefully, for they stunningly illustrate how a great scientific mind can grasp these eternal truths revealed in the Bible:

Wandering and weak are all our prayers,
And fleeting half the gifts we crave;
Love only, cleansed from sins and care,
Shall live beyond the grave.

Strengthen our love, O Lord, that we
May in Thine own great love believe
And, opening all our soul to Thee,
May Thy free gift receive.

All powers of mind, all force of will,
May lie in dust when we are dead,
But love is ours, and shall be still,
When earth and seas are fled.[2556]

APPENDIX

DETAILS ON FREQUENTLY CITED SOURCES

ANCIENT WORK: MISHNAH

(Translated and introduced by Herbert Danby. Oxford University Press, 1954. First published in 1933.)

The introduction states the Mishnah is a Jewish holy book primarily concerned with religious laws. Most of this work was compiled and edited around 200 A.D. by the spiritual leader of Judaism at that time (Rabbi Judah). The "essential and characteristic element" of the Mishnah is that it contains "the Oral Law" of that time. Generally speaking, the Oral Law is comprised of "beliefs and religious practices" that are not explicitly written in the Hebrew Scriptures.

On page xii of the introduction, Danby states that after the destruction of the Temple in 70 A.D., the sect called the Pharisees "took the position, naturally and almost immediately, of sole and undisputed leaders of such Jewish life as survived. Judaism as it has continued since is, if not their creation, at least a faith and a religious institution largely of their fashioning; and the Mishnah is the authoritative record of their labor." Page xv states the Mishnah has a "purely Pharisean outlook."

Another modern English translation of the Mishnah (by Jacob Neusner. Yale University Press, 1988) takes issue with this viewpoint. On pages xxxii–iii, Neusner argues the Mishnah "cannot be called a document only or mainly" of the Pharisees. In support of this view, he writes:

> [T]he Mishnah rarely refers to the Pharisees. When it does, it does not represent them as definitive authorities. Sages, not Pharisees, are the Mishnah's authorities. A few of the Mishnah's authorities, particularly Gamaliel and Simeon b. Gamaliel, are known from independent sources to have been Pharisees; Paul tells us about Gamaliel, and Josephus about Simeon b. Gamaliel. But that is the sum and substance of it. Consequently, to assign the whole of the Mishnah to the Pharisees who flourished before

A.D. 70 and who are known to us from diverse sources, all of them composed in the form in which we know them after A.D. 70, is hardly justified.

With tremendous respect for Jacob Neusner, whose scholarly contributions are invaluable to the study of ancient Judaism, I find Danby's view more sustainable. Although Josephus writes about the Pharisees in works that were published after 70 A.D., he was born in 37 A.D. and was well ensconced in public affairs before the Temple was destroyed.[2557] Furthermore, in his autobiography, Josephus writes that he "began to conduct" himself "according to the rules of the sect of the Pharisees" at the age of nineteen, which was 56 A.D.[2558] Given these facts and the scarcity of Jewish historical works from this era, one cannot reasonably expect to find a more credible source concerning the Pharisees. I would also dispute that all of Paul's writings were composed after 70 A.D. However, we don't need to delve into dating Paul's writings because the writings of Josephus go a long way towards resolving this issue. In the *Antiquities*, Josephus writes:

> [T]he Pharisees have delivered to the people a great many observances by succession from their fathers, which are not written in the laws of Moses [the first five books of the Bible]; and for that reason it is that the Sadducees [an opposing sect] reject them, and say that we are to esteem those observances to be obligatory which are in the written word, but are not to observe what are derived from the tradition of our forefathers. And concerning these things it is that great disputes and differences have arisen among them....[2559]

To repeat, the primary defining element of the Pharisees is that they "delivered to the people a great many observances by succession from their fathers, which are not written in the laws of Moses." Open to just about any page in the Mishnah and that is precisely what you'll find. The Mishnah is a compilation of exactly what Josephus said the Pharisees followed: a tremendous number of laws not found in the Jewish scriptures along with an account of how they were passed down from generation to generation.[2560]

That the Mishnah rarely uses the word "Pharisee" can be easily explained by the fact that it was compiled generations after the destruction of the Temple. If the doctrines of the Pharisees were ubiquitous at this time and the term "Pharisee" fell out of use because they were no longer considered a sect but represented most all of Judaism, then it would make perfect sense for the authorities in the Mishnah to be referred to as sages. The following quote from the *Jewish Encyclopedia* fits perfectly with this scenario:

> [W]ith the destruction of the Temple, the Sadducees disappeared altogether, leaving the regulation of all Jewish affairs in the hands of the Pharisees. Henceforth Jewish life was regulated by the teachings of the Pharisees…. Pharisaism shaped the character of Judaism and the life and thought of the Jew for all the future.[2561]

Also in concert with this is the following passage from the Mishnah. One of the primary authorities in it is a rabbi by the name of Johanan ben Zakkai. He was the most important sage of the periods before, during, and just after the destruction of the Temple.[2562] The Mishnah quotes him as stating, "Have we naught against the Pharisees save this!"[2563] The Mishnah then goes on to describe a disagreement that Johanan had with a Pharisaic doctrine. What this tells us is that Johanan agreed with the teachings of the Pharisees on everything but this one issue. Given that the Mishnah contains thousands of laws and that it identifies Johanan ben Zakkai as the keeper of "the Law" during his generation,[2564] this is an extremely significant statement.

Further supporting the view that the doctrines of the Pharisees became the norm after the destruction of the Temple is the fact that the Mishnah cites rules attributed to Gamaliel and Simeon ben Gamaliel without even hinting that they were Pharisees.[2565] [2566]

In summary, by the time the Mishnah was compiled and edited around 200 A.D., the label "Pharisee" had clearly fallen out of usage. However, the doctrines found in the Mishnah generally reflect those of a people, known as the Pharisees, from an earlier era.

Ancient Work: Tosefta

(Translated by Jacob Neusner. KTAV Publishing House.)

This translation is composed of six divisions published between 1977 and 1986. Page ix in the introduction to the first division states, "The Tosefta, a collection of statements to supplement the Mishnah, came to closure about two centuries after the Mishnah, one may guess at about 400." It has the same basic format as the Mishnah, and the "authorities" cited in it are the same authorities cited in the Mishnah.

Ancient Work: Babylonian Talmud

(Multiple translations and editions.)

The Babylonian Talmud is a series of commentaries pertaining to Jewish civil and religious laws. It contains excerpts from the Mishnah, surrounded by interpretations and supplemental information provided by Jewish scholars who resided in Babylonia between the 3rd and 6th centuries A.D. When the word "Talmud" is used by itself, this commonly refers to the Babylonian Talmud as opposed to the Talmud of the Land of Israel (see below), because it is the more popular of the two.

Ancient Work: Talmud of the Land of Israel

(Multiple translations and editions.)

Also called the Jerusalem Talmud, Palestinian Talmud, or Talmud Yerushalmi. The Jerusalem Talmud is a series of commentaries pertaining to Jewish civil and religious laws. It contains excerpts from the Mishnah, surrounded by interpretations and supplemental information provided by Jewish scholars who resided in the land of Israel between the 3rd and 6th centuries A.D.

Citations and index located at
www.RationalConclusions.com and
in e-reader versions of this book.

Printed and bound by PG in the USA